Biotechnology and Its Applications in Agricultural Science

Biotechnology and Its Applications in Agricultural Science

Anand Prakash

RANDOM

RANDOM PUBLICATIONS

NEW DELHI (INDIA)

Biotechnology and Its Applications in Agricultural Science

ISBN 978-93-5111-209-9

Published in 2014 in India by

RANDOM PUBLICATIONS

4376-A/4B, Gali Murari Lal, Ansari Road
New Delhi-110 002
Phone : +91-11-43580356, +91-11-23289044
e-mail: randomexports@gmail.com, sales@randompublications.com,
info@randompublications.com

Reprinted 2026

Type Setting by : Keystoneprintads, Delhi-110051

Preface

Biotechnology is broadly defined as any technique that uses live organisms viz. bacteria, viruses, fungi, yeast, animal cells, plant cells etc. to make or modify a product, to improve plants or animals or to engineer micro-organisms for specific uses.

It encompasses genetic engineering, inclusive of enzyme and protein engineering plant and animal tissue culture technology, biosensors for biological monitoring, bioprocess and fermentation technology.

Biotechnology is essentially and interdisciplinary are consisting of biochemistry, molecular chemistry, molecular and microbiology, genetics and immunology etc. it is concerned with upgradation of quality and also utilization of livestock and resources for the well being of both animals and plants. Biotechnology is any technique that uses living organisms or substances from these organisms to make or modify a product for a practical purpose.

Biotechnology can be applied to all classes of organism - from viruses and bacteria to plants and animals - and it is becoming a major feature of modern medicine, agriculture and industry. Modern agricultural biotechnology includes a range of tools that scientists employ to understand and manipulate the genetic make-up of organisms for use in the production or processing of agricultural products.

The development of genetically modified foods and other agricultural biotechnology products has generated significant public debate. The potential for creating foods enhanced for health benefits or increasing crop yields was tantalizing, but there was also widespread concern about the technology's health and environmental risks.

The book intends to explain the basic aspects of agricultural biotechnology in a simple and lucid manner.

I thank all members of my team who have helped in the preparation of the book. My special thanks go to "Random Publications" who have published the book.

– ***Anand Prakash***

Contents

1

Agricultural Biotechnology

Broadly speaking, biotechnology is any technique that uses living organisms or substances from these organisms to make or modify a product for a practical purpose. Biotechnology can be applied to all classes of organism - from viruses and bacteria to plants and animals - and it is becoming a major feature of modern medicine, agriculture and industry. Modern agricultural biotechnology includes a range of tools that scientists employ to understand and manipulate the genetic make-up of organisms for use in the production or processing of agricultural products.

Some applications of biotechnology, such as fermentation and brewing, have been used for millennia. Other applications are newer but also well established. For example, micro-organisms have been used for decades as living factories for the production of life-saving antibiotics including penicillin, from the fungus *Penicillium*, and streptomycin from the bacterium *Streptomyces*. Modern detergents rely on enzymes produced via biotechnology, hard cheese production largely relies on rennet produced by biotech yeast and human insulin for diabetics is now produced using biotechnology.

Biotechnology is being used to address problems in all areas of agricultural production and processing. This includes plant breeding to raise and stabilize yields; to improve resistance to pests, diseases and abiotic stresses such as drought and cold; and to enhance the nutritional content of foods. Biotechnology is being used to develop low-cost disease-free planting materials for crops such as cassava, banana and potato and is creating new tools for the diagnosis and treatment of plant and animal diseases and for the measurement and conservation of genetic resources. Biotechnology is being used to speed up breeding programmes for plants, livestock and fish and to extend the range of traits that can be addressed. Animal feeds and feeding practices are being changed by biotechnology to improve animal nutrition and to reduce environmental waste. Biotechnology is used in disease diagnostics and for the production of vaccines against animal diseases.

Clearly, biotechnology is more than genetic engineering. Indeed, some of the least controversial aspects of agricultural biotechnology are potentially the most powerful and the most beneficial for the poor. Genomics, for example, is revolutionizing our understanding of the ways genes, cells, organisms and ecosystems function and is opening new horizons for marker-assisted breeding and genetic resource management. At the same time, genetic engineering is a very powerful tool whose role should be carefully evaluated. It is important to understand how biotechnology - particularly genetic engineering - complements and extends other approaches if sensible decisions are to be made about its use. This chapter provides a brief description of current and emerging uses of biotechnology in crops, livestock, fisheries and forestry with a view to understanding the technologies themselves and the ways they complement and extend other approaches. It should be emphasized that the tools of biotechnology are just that: tools, not ends in themselves. As with any tool, they must be assessed within the context in which they are being used.

The Convention on Biological Diversity (CBD) defines biotechnology as: "any technological application that uses biological systems, living organisms, or derivatives thereof, to make or modify products for specific use". This definition includes medical and industrial applications as well as many of the tools and techniques that are commonplace in agriculture and food production.

The Cartagena Protocol on Biosafety defines "modern biotechnology" more narrowly as the application of:

- In vitro nucleic acid techniques, including recombinant deoxyribonucleic acid (DNA) and direct injection of nucleic acid into cells or organelles, or
- Fusion of cells beyond the taxonomic family, that overcome natural physiological reproductive or recombination barriers and that are not techniques used in traditional breeding and selection.

Recombinant DNA techniques, also known as genetic engineering or (more familiarly but less accurately) genetic modification, refer to the modification of an organism's genetic make-up using transgenesis, in which DNA from one organism or cell (the transgene) is transferred to another without sexual reproduction. Genetically modified organisms (GMOs) are modified by the application of transgenesis or recombinant DNA technology, in which a transgene is incorporated into the host genome or a gene in the host is modified to change its level of expression. The terms "GMO", "transgenic organism" and "genetically engineered organism (GEO)" are often used interchangeably although they are not technically identical. For the purposes of this report they are used as synonyms.

UNDERSTANDING, CHARACTERIZING AND MANAGING GENETIC RESOURCES

Farmers and pastoralists have manipulated the genetic make-up of plants and animals since agriculture began more than 10 000 years ago. Farmers managed the process of domestication over millennia, through many cycles of selection of the best adapted individuals. This exploitation of the natural variation in biological organisms has given us the crops, plantation trees, farm animals and farmed fish of today, which often differ radically from their early ancestors. The aim of modern breeders is the same as that of early farmers - to produce superior crops or animals.

Conventional breeding, relying on the application of classic genetic principles based on the phenotype or physical characteristics of the organism concerned, has been very successful in introducing desirable traits into crop cultivars or livestock breeds from domesticated or wild relatives or mutants. In a conventional cross, whereby each parent donates half the genetic make-up of the progeny, undesirable traits may be passed on along with the desirable ones, and these undesirable traits may then have to be eliminated through successive generations of breeding. With each generation, the progeny must be tested for its growth characteristics as well as its nutritional and processing traits.

Table. An Agricultural Technology Timeline

Technology	Era	Genetic Interventions
Traditional	About 10 000 years BC	Civilizations harvested from natural biological diversity, domesticated crops and animals, began to select plant materials for propagation and animals for breeding
	About 3 000 years BC	Beer brewing, cheese making and wine fermentation
Conventional	Late nineteenth century	Identification of principles of inheritance by Gregor Mendel in 1865, laying the foundation for classical breeding methods
	1930s	Development of commercial hybrid crops
	1940s to 1960s	Use of mutagenesis, tissue culture, plant regeneration. Discovery of transformation and transduction. Discovery by Watson and Crick of the structure of DNA in 1953. Identification of genes that detach and move (transposons)
Modern	1970s	Advent of gene transfer through recombinant DNA techniques. Use of embryo rescue and protoplast fusion in plant breeding and artificial insemination in animal reproduction
	1980s	Insulin as first commercial product from gene transfer. Tissue culture for mass propagation in plants and embryo transfer in animal production
	1990s	Extensive genetic fingerprinting of a wide range of organisms. First field trials of genetically engineered plant varieties in 1990 followed by the first commercial release in 1992. Genetically engineered vaccines and hormones and cloning of animals
	2000s	Bioinformatics, genomics, proteomics, metabolomics

Many generations may be required before the desired combination of traits is found, and time lags may be very long, especially for perennial crops such as trees and some species of livestock. Such phenotype-based selection is thus a slow, demanding process and is expensive in terms of both time and money. Biotechnology can make the application of conventional breeding methods more efficient.

Spontaneous mutations are the "natural" motor of evolution, and the resource into which breeders tap to domesticate crops and to "create" better varieties. Without mutations, there would be no rice, or maize or any other crop. Starting in the 1970s, the International Atomic Energy Agency (IAEA) and FAO sponsored research on mutation induction to enhance genetic improvement of food and industrial crops for breeding new improved varieties. Induced mutations are brought about by treating plant parts with chemical or physical mutagens and then selecting for desirable changes - in effect, to mimic spontaneous mutations and artificially broaden genetic diversity. The precise nature of the mutations induced has generally not been a concern irrespective of whether the mutant lines were used directly or as sources of new variation in cross-breeding programmes.

Induced mutation to assist breeding has resulted in the introduction of new varieties of many crops such as rice, wheat, barley, apples, citrus, sugar cane and banana (the FAO/IAEA Mutant Varieties Database lists more than 2 300 officially released varieties). The application of mutation induction to crop breeding has translated into a tremendous economic impact on agriculture and food production that is currently valued in billions of US dollars and millions of hectares of cultivated land. Recently, mutation techniques have undergone a renaissance, expanding beyond their direct use in breeding into novel applications such as gene discovery and reverse genetics.

GENOMICS

The most significant breakthroughs in agricultural biotechnology are coming from research into the structure of genomes and the genetic mechanisms behind economically important traits. The rapidly progressing discipline of genomics is providing information on the identity, location, impact and function of genes affecting such traits - knowledge that will increasingly drive the application of biotechnology in all agricultural sectors. Genomics sets the foundation for post-genomics activities, including new disciplines such as proteomics and metabolomics to generate knowledge on gene and protein structure, as well as their functions and interactions. These disciplines seek to understand systematically the molecular biology of organisms for their practical use.

A vast range of new and rapidly advancing technologies and equipment has also been developed to generate and process information about the structure and function of biological systems. The use and organization of this information is called bioinformatics. Advances in bioinformatics may allow the prediction of gene function from gene sequence data: from a listing of an organism's genes, it will become possible to build a theoretical framework of its biology. The comparison across organisms of physical and genetic maps and DNA sequences will significantly reduce the time needed to identify and select potentially useful genes.

Through the production of genetic maps that provide the precise location and sequences of genes, it is apparent that even distantly related genomes share common features. Comparative genomics assists in the understanding of many genomes based on the intensive study of just a few. For instance, the rice genome sequence is useful for studying the genomes of other cereals with which it shares features according to its degree of relatedness, and the mouse and malaria genomes provide models for livestock and some of the diseases that affect them. There are now model species for most types of crops, livestock and diseases and knowledge of their genomes is accumulating rapidly.

DNA

All living things are made up of cells that are programmed by genetic material called deoxyribonucleic acid (DNA). Only a small fraction of the DNA chain actually makes up genes, which in turn code for proteins, and the remaining share of the DNA represents non-coding sequences whose role is not yet clearly understood. The genetic material is organized into pairs of chromosomes. For example, there are five chromosome pairs in the much-studied mustard species Arabidopsis thaliana. An organism's entire set of chromosomes is called the genome. The Human Genome Sequencing Project has provided the agricultural research community not only with many spin-off technologies that can be applied across the board for all living organisms but also with a model for international collaboration in tackling large genome-sequencing projects for model plants such as Arabidopsis and rice.

Synteny describes the conservation or consistency of gene content and gene order along the chromosomes of different plant genomes. Until well into the 1980s we imagined that each crop plant had its own genetic map. Only when we were able to make the first molecular maps, using a technique called "restriction fragment length polymorphism" (RFLP), did it begin to dawn on us that related species had remarkably similar gene maps.

The early experiments demonstrated conservation over a few million years of evolution in syntenous relationships between potato and tomato in the broad-leafed plants and between the three genomes of bread wheat in the grasses. Later we were able to show that the same similarities held over the rice, wheat and maize genomes, which were separated by some 60 million years of evolution. The diagram summarizes this research and shows 70 per cent of the world's food linked in a single map. The 12 chromosomes of rice can be aligned with the ten chromosomes of maize and the basic seven chromosomes of wheat and barley in such a way that any radius drawn around the circles will pass through different versions, known as alleles, of the same genes.

The discovery of synteny has had an enormous impact on the way we think about plant genetics. There are obvious applications for evolutionary studies; for example, the white arrows on the wheat and maize circles describe evolutionary chromosomal translocations that describe *Pooideae* and *Panicoideae* groups of grasses. There are great opportunities to predict the presence and location of a gene in one species from what we know from another. Now that we have the complete DNA sequence of rice we are able to identify and isolate key genes from large genome intractable species such as wheat and barley by predicting that the same genes will be present in the same order as in rice. Key genes for disease resistance and tolerance to acid soils have recently been isolated from barley and rye in this way. For practical plant breeding, knowledge of synteny allows breeders access to all alleles in, for example, all cereals rather than just the species on which they are working. A key first example of this is the transfer to rice of the wheat dwarfing genes that made the Green Revolution possible. In these experiments the gene was located in rice by synteny and then isolated and engineered with the alteration in DNA sequence that characterized the wheat genes before replacing the engineered gene in rice. This approach can be applied to any gene in any cereal, including the so-called "orphan crops" that have not attracted the research dollars that the big three - wheat, rice and maize - have over the past century. The main significance is, however, that we can now pool our knowledge of biochemistry, physiology and genetics and transfer it between crops via synteny.

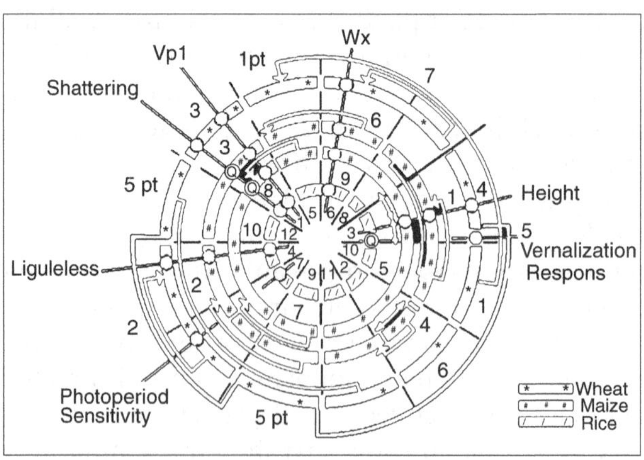

MOLECULAR MARKERS

Reliable information on the distribution of genetic variation is a prerequisite for sound selection, breeding and conservation programmes. Genetic variation of a species or population can be assessed in the field or by studying molecular and other markers in the laboratory. A combination of the two approaches is required for reliable results. Molecular markers are identifiable DNA sequences, found at specific locations of the genome and associated with the inheritance of a trait or linked gene.

Molecular markers can be used for:

- Marker-assisted breeding,
- Understanding and conserving genetic resources
- Genotype verification. These activities are critical for the genetic improvement of crops, forest trees, livestock and fish.

MARKER-ASSISTED BREEDING

Genetic linkage maps can be used to locate and select for genes affecting traits of economic importance in plants or animals. The potential benefits of marker-assisted selection (MAS) are greatest for traits that are controlled by many genes, such as fruit yield, wood quality, disease resistance, milk and meat production, or body fat, and that are difficult, time-consuming or expensive to measure. Markers can also be used to increase the speed or efficiency of introducing new genes from one population to another, for example when wishing to introduce genes from wild relatives into modern plant varieties. When the desired trait is found within the same species, it may be transferred with traditional breeding methods, with molecular markers being used to track the desired gene.

Pearl millet is a cereal grown for foodgrain and straw in the hottest, driest areas of Africa and Asia where rainfed and dryland agriculture are practised. It is similar to maize in its breeding behaviour. Traditional farmers' varieties are open-pollinated and out-breeding and thus continuously changing. Genetically uniform hybrid varieties have been developed that offer higher yield potential but are more vulnerable to a plant disease called downy mildew.

In India, pearl millet is grown on about 9 million ha and more than 70 per cent of this is sown to such hybrid cultivars. Since pearl millet hybrids first reached farmers' fields in India in the late 1960s, every variety that has become popular with farmers has ultimately succumbed to a downy mildew epidemic. Unfortunately, by the time the poorer farmers in a given region decide to adopt a particular variety, its days are usually numbered.

The International Crops Research Institute for the Semi-Arid Tropics (ICRISAT) wanted to reduce the risks associated with adoption of higher-yielding pearl millet hybrids and extend the useful economic life of these varieties, especially for poorer producers. Biotechnology helped us to achieve this. With tools from the John Innes Centre and support from the Plant Sciences Research Programme of the Department for International Development (DFID), we developed and applied molecular genetic tools for pearl millet. We mapped the genomic regions of pearl millet that control downy mildew resistance, straw yield potential, and grain and straw yield under drought stress conditions. Then our millet breeders used conventional breeding and marker-assisted selection (MAS) to transfer several genomic regions conferring improved downy mildew resistance to the two elite inbred parental lines of popular hybrid HHB 67. We then used MAS to derive two new varieties - ICMR 01004 and ICMR 01007 - with two different downy mildew resistance gene blocks.

These varieties have performed as well or better than their parent lines for grain and straw yield, and are markedly improved for downy mildew resistance. They also retain several favourable traits, including 1 000-grain mass, panicle length, plant height and rust resistance. Hybrids based on crosses involving ICMR 01004 and ICMR 01007 have recently advanced to trials in the Indian states of Gujarat, Rajasthan and Haryana under the All India Coordinated Pearl Millet Improvement Project. This follows their successful evaluation in 2002, in which they exhibited marginal grain yield superiority and substantially better downy mildew resistance than HHB 67, while maintaining the early maturity that contributes to its popularity.

At least one of these two hybrids could be released as a replacement for HHB 67 before the latter succumbs (as it surely will) to a downy mildew epidemic. Because HHB 67 is so widely grown by poor farmers in India, if its timely replacement could prevent such an epidemic for even one year, the losses avoided would exceed the total value of research-funding support by DFID for the development and application of the molecular genetic tool kit for pearl millet (£3.1 million to date). All future benefits from this research by ICRISAT, its DFID-supported partners in the United Kingdom, and collaborating national programme partners in India can then be considered profits to society.

MEASURING AND CONSERVING GENETIC DIVERSITY

The use of molecular markers to measure the extent of variation at the genetic level, within and among populations, is of value in guiding genetic conservation activities and in the development of breeding populations in crops, livestock, forestry and fisheries. Studies carried out using these technologies in fish and forest tree species have revealed high levels of genetic variation both among and within populations. Livestock species are characterized by a high degree of genetic variation within populations, whereas crops exhibit a higher degree of variation across species. Data from other approaches, for example field observation, often cannot provide such information or are extremely difficult to collect.

Molecular markers are increasingly used to study the distribution and patterns of genetic diversity. Global surveys indicate, for example, that 40 per cent of the remaining domestic livestock breeds are at risk of extinction. Most of these breeds are found only in developing countries, and there is often little knowledge about them or of their potential for improvement. They may contain valuable genes that confer adaptation or resilience to stresses, such as heat tolerance or disease resistance, that may be of use for future generations. Modern biotechnologies can help to counteract trends of genetic erosion in all food and agriculture sectors.

SIGNIFICANT OF AGRICULTURAL BIOTECHNOLOGY

Every year, millions of people worldwide die from starvation and nutritional deficiencies. Although these challenges remain significant, we are making some progress. Agricultural technologies have increased caloric intake on a per-capita basis in nearly every nation of the world. technological development— particularly in the area of agricultural biotechnology — is crucial if we want to further reduce malnutrition and starvation and meet the needs of growing populations. Unfortunately, misperceptions about biotechnology have led to activist movements to slow down, if not eliminate, this critical technology via government regulation. In the United States, the advancement of biotech crops has slowed down because of attempts to regulate them differently than conventionally grown foods. In Europe, antibiotech movements are much stronger, and international bureaucrats are pushing various international agreements and trade restrictions. While Western nations may be able to absorb the costs of such policies, people in the developing world will pay a higher price as many continue to suffer from starvation.

WHAT IS AGRICULTURAL BIOTECHNOLOGY?

Broadly speaking, biotechnology is any technique that uses living organisms or substances from these organisms to make or modify a product for a practical purpose. Biotechnology can be applied to all classes of organism-from viruses and bacteria to plants and animals-and it is becoming a major feature of modern medicine, agriculture and industry. Modern agricultural biotechnology includes a range of tools that scientists employ to understand and manipulate the genetic make-up of organisms for use in the production or processing of agricultural products.

Some applications of biotechnology, such as fermentation and brewing, have been used for millennia. Other applications are newer but also well established. For example, micro-organisms have been used for decades as living factories for the production of life-saving antibiotics including penicillin, from the fungus Penicillium, and streptomycin from the bacterium Streptomyces. Modern detergents rely on enzymes produced via biotechnology, hard cheese production largely relies on rennet produced by biotech yeast and human insulin for diabetics is now produced using biotechnology.

Biotechnology is being used to address problems in all areas of agricultural production and processing. This includes plant breeding to raise and stabilize yields; to improve resistance to pests, diseases and abiotic stresses such as drought and cold; and to enhance the nutritional content of foods. Biotechnology is being used to develop low-cost disease-free planting materials for crops such as cassava, banana and potato and is creating new tools for the diagnosis and treatment of plant and animal diseases and for the measurement and conservation of genetic resources. Biotechnology is being used to speed up breeding programmes for plants, livestock and fish and to extend the range of traits that can be addressed. Animal feeds and feeding practices are being changed by biotechnology to improve animal nutrition and to reduce environmental waste. Biotechnology is used in disease diagnostics and for the production of vaccines against animal diseases.

Clearly, biotechnology is more than genetic engineering. Indeed, some of the least controversial aspects of agricultural biotechnology are potentially the most powerful and the most beneficial for the poor. Genomics, for example, is revolutionizing our understanding of the ways genes, cells, organisms and ecosystems function and is opening new horizons for marker-assisted breeding and genetic resource management. At the same time, genetic engineering is a very powerful tool whose role should be carefully evaluated. It is important to understand how biotechnology-particularly genetic engineering-complements and extends other approaches if sensible decisions are to be made about its use.

UNDERSTANDING, CHARACTERIZING AND MANAGING GENETIC RESOURCES

Farmers and pastoralists have manipulated the genetic make-up of plants and animals since agriculture began more than 10 000 years ago. Farmers managed the process of domestication over millennia, through many cycles of selection of the best adapted individuals. This exploitation of the natural variation in biological organisms has given us the crops, plantation trees, farm animals and farmed fish of today, which often differ radically from their early ancestors.

The aim of modern breeders is the same as that of early farmers-to produce superior crops or animals. Conventional breeding, relying on the application of classic genetic principles based on the phenotype or physical characteristics of the organism concerned, has been very successful in introducing desirable traits into crop cultivars or livestock breeds from domesticated or wild relatives or mutants. In a conventional cross, whereby each parent donates half the genetic make-up of the progeny, undesirable traits may be passed on along with the desirable ones, and these undesirable traits may then have to be eliminated through successive generations of breeding. With each generation, the progeny must be tested for its growth characteristics as well as its nutritional and processing traits. Many generations may be required before the desired combination of traits is found, and time lags may be very long, especially for perennial crops such as trees and some species of livestock. Such phenotype-based selection is thus a slow, demanding process and is expensive in terms of both time and money. Biotechnology can make the application of conventional breeding methods more efficient.

GENOMICS

The most significant breakthroughs in agricultural biotechnology are coming from research into the structure of genomes and the genetic mechanisms behind economically important traits. The rapidly progressing discipline of genomics is providing information on the identity, location, impact and function of genes affecting such traits-knowledge that will increasingly drive the application of biotechnology in all agricultural sectors. Genomics sets the foundation for post-genomics activities, including new disciplines such as proteomics and metabolomics to generate knowledge on gene and protein structure, as well as their functions and interactions. These disciplines seek to understand systematically the molecular biology of organisms for their practical use.

A vast range of new and rapidly advancing technologies and equipment has also been developed to generate and process information about the structure and function of biological systems. The use and organization of this information is called bioinformatics. Advances in bioinformatics may allow the prediction of gene function from gene sequence data: from a listing of an organism's genes, it will become possible to build a theoretical framework of its biology. The comparison across organisms of physical and genetic maps and DNA sequences will significantly reduce the time needed to identify and select potentially useful genes.

Through the production of genetic maps that provide the precise location and sequences of genes, it is apparent that even distantly related genomes share common features. Comparative genomics assists in the understanding of many genomes based on the intensive study of just a few.

For instance, the rice genome sequence is useful for studying the genomes of other cereals with which it shares features according to its degree of relatedness, and the mouse and malaria genomes provide models for livestock and some of the diseases that affect them. There are now model species for most types of crops, livestock and diseases and knowledge of their genomes is accumulating rapidly.

BIOTECHNOLOGICAL PRINCIPAL OF AGRICULTURE

Agricultural technology can be defined as biotechnology to include 'any technique that uses living organisms or processes to make or modify products, to improve plants or animals or to develop micro-organisms for specific uses-it focuses upon recombinant DNA and cell fusion technologies'. Longworth concurs with this definition, with the significant addition of tissue-culture techniques, an aspect of biotechnology which has great economic, and therefore social and political, potential impact.

Despite debates about whether these or any other definitions are adequate, they do touch on the main aspects of biotechnology which deserve to be considered. The aspect of this new biotechnology which most captures the imagination and stirs the greatest controversy is gene splicing or recombinant DNA techniques, which inspire researchers to consider the possibilities of producing reproducible animals and plants markedly different from current existing species, and referred to as transgenic species. Already in higher animals and plants recombinant DNA has produced transgenic forms which are being commercially exploited. Their agricultural significance is, however, so far limited, and commercial application is concentrated in highly profitable pharmaceutical and horticultural markets.

Genes have been introduced into several animal species which alter their protein synthesis to enable transgenic sheep to produce insulin in their milk, and rabbits to produce interferon. A completely different application has originated in Denmark for salmon, where it has proved possible to introduce germplasm which enables the salmon's physiology to handle heavy metals, which are normally toxic, so opening up new locations for farm fisheries. Much current research is directed to conferring disease immunity on animals, and holds out the prospect of widespread commercial application.

In plants one achievement has been the transfer of genetic resistance to antibiotics in the petunia, and another has been the introduction of storage-protein genes from French bean plants into tobacco plants. As yet commercial progress with recombinant-DNA technology in plants appears limited, but extensive opportunities beckon, particularly because plant research is less restricted by the ethical and animal-welfare concerns which apply to research on transgenic animals. The most important commercial developments based on gene splicing have so far occurred with genetically much simpler microorganisms, and it is with these that the greatest short-to mediumrun commercial potential lies.

Already genetically engineered micro-organisms are producing a variety of hormones, vaccines, enzymes and other proteins. Important examples are the production of insulin, vaccines for neo-natal diarrhoea in calves and piglets, and the bovine growth hormone BST identical to that produced in cows which can stimulate a 20 per cent increase in milk yield. BST is already a source of problems for legislators in the European Community, and has provoked strong reactions from the media and milk consumers. As far as larger animals, and cattle in particular, are concerned, it is developments in embryo transfer and in many processes for manipulating reproduction which hold out the prospect of continuing increases in yields, feed conversion efficiency and general economic efficiency. Already, apparently over 1 per cent of dairy calves in the USA are from embryo transplants, despite the high costs still associated with this procedure. Widespread adoption of these sophisticated technologies would further distance livestock farming from its traditional rural simplicity, and from the natural mating of animals to produce offspring. It would place technical demands upon the operators which favour large-scale company farms capable of supporting a range of highly trained specialists. Longworth identifies new techniques of tissue and cell culture as having 'the potential for enormous advances in crop improvement in the next couple of decades'. Those techniques 'can both increase the genetic diversity and greatly increase selection efficiency', and they permit innumerable plants to be reproduced asexually from single cells or small pieces of tissue.

The particular technique known as 'callus culture' has been used for many years to clone highly valued horticultural plants such as orchids, and cloning of cuttings is widely practised for tree crops such as tea and palm oil as well as by millions of gardeners for garden plants. According to Longworth 'cell culture' of single cells has unexpectedly, and so far inexplicably, resulted in plants with different properties being regenerated from the same clump of parent tissue. Sugar cane, maize and potato plants regenerated in this way have been found which are resistant to important pathogens.

This application of cell culture with the capacity to generate vast numbers of seedlings rapidly has the potential to simplify greatly the hitherto labourious procedures of plant breeding and selection. However, in terms of current and immediate commercial importance, it is through micro-organisms that biotechnology has its greatest impact. Microbial fermentation processes have been of great commercial significance for centuries, for example in bread, wine and cheese production, and there is the prospect of considerable development. Two recent examples of new processes indicate the sorts of impact that such developments can have.

In the late 1960s genetically engineered bacteria were developed which were able to digest corn starch to produce high-fructose corn syrup (HFCS) and which left as a residue corn gluten, which is now an important protein feed for livestock. HFCS has made considerable inroads in the United States and in Japan. It has largely replaced sugar as a sweetener in Coca Cola as well as many other food and drink products. This has helped depress sugar prices to Third World growers. In the European Community steps have been taken to prevent HFCS and other new powerful sweeteners from being produced and from undermining the market for domestically grown beet sugar, which is supported by the Common Agricultural Policy.

A second important application of microbial fermentation has been the production of ethanol from sugar cane in Brazil and from maize in the USA. The commercial viability of these processes is critically dependent upon the price of oil as the main non-renewable source of fuel, and, to date, massive subsidies have been required to maintain the Brazilian and USA ethanol programmes. Longworth states, however, that a new biotechnology, Sucrotech, is being patented, which will not only reduce the cost of producing ethanol from sugar cane, but will simultaneously produce fructose at a cost which will be competitive with HFCS and may thereby reclaim part of the sweetener market for sugar cane.

That such possibilities are in prospect indicates how volatile the future might be; biotechnology has tilted the competitive balance from sugar cane to maize, causing economic pressure and even disruption to sugar-cane-dependent economies and may in future switch it back again. In the long run the capacity to ferment fuels microbially from renewable agricultural feedstocks points to an important long-term reorientation of agriculture if non-renewable oil becomes uncompetitively expensive as a fuel for cars and feedstock for certain chemicals.

It suggests that eventually there will be an increased emphasis on agricultural production of industrial feedstocks at the same time as continually increasing food output will be required to feed the expanding world population. All this will require considerable increases in agricultural productivity, to which biotechnology will increasingly contribute, but it will at the same time impose great strains on the natural environment and the structure of agriculture.

AGRICULTURE ON ACID SOILS

Aluminium in acid soils limits plant growth on more than 30 per cent of all arable land, primarily in developing countries. There are two approaches to increasing crop production on acid soils. Lime can be added to the soil to increase the pH, but this is a costly, temporary measure. Alternatively, genetically improved cultivars, tolerant to aluminium, can be developed. Existing wheat cultivars do not contain significant genetic variation for increasing aluminium tolerance. Improved tolerance will have to be introduced into wheat from the gene pools of related, more tolerant species. A genetic linkage map of wheat was developed using available markers for the existing aluminium-tolerance gene. Rye exhibits a fourfold increase in aluminium tolerance over wheat. Therefore, a rye gene controlling aluminium tolerance was characterized. Markers from wheat, barley and rice were used to establish a tight linkage, flanking the rye gene, and to construct a high-resolution genetic map. A potential candidate gene was used for root-gene-expression, time-course studies that showed expression in rye roots only under aluminium stress.

Targeting the aluminium tolerance gene is one example of using problem-based approaches to integrate molecular and breeding tools to improve wheat production. Using the genetic relationship (synteny) among the cereals to supply markers to identify and characterize value-added traits, complementary approaches for improved wheat production emerge. Breeders can use the markers flanking the rye gene in marker-assisted breeding programmes in areas where GMOs cannot be grown or where only conventional breeding tools are available. In addition, these markers can be used for map-based cloning to isolate the gene in question for transgenic approaches to wheat improvement. Finally, the use of syntenous relationships offers the technology to manipulate many value-added traits for crop improvement in other species.

RESEARCH

Researchers at Jawaharlal Nehru University in India have developed a genetically engineered potato that produces about one-third to one-half more protein than usual, including substantial amounts of all the essential amino acids such as lysine and methionine. Protein deficiency is widespread in India and potato is the staple food of the poorest people.

The "protato" was developed by a coalition of Indian charities, scientists, government institutes and industry as part of a 15-year campaign against childhood mortality. The campaign aims to eliminate childhood mortality by providing children with clean water, better food and vaccines. The protato includes a gene from the amaranth plant, a high-protein grain that is native to South America and widely sold in Western health-food stores. The protato has passed preliminary field trials and tests for allergens and toxins. Final approval from the Indian Government is probably at least five years away.

Supporters such as Govindarajan Padmanaban, a biochemist at the Indian Institute of Science, argue that the protato can provide an important nutritional boost to children with little danger of allergy because potatoes and amaranth are both already widely consumed. There is also little threat to the environment because neither potatoes nor amaranth have wild relatives in India, and the protato does not involve any change in normal potato production practices.

Furthermore, because the protato was developed by public-sector scientists in India, there are no concerns about foreign corporate control of the technology.

Given these benefits, Padmanaban commented:

- "I think it would be morally indefensible to oppose it".

Opponents such as Charlie Kronick of Greenpeace argue that potatoes are naturally quite low in protein (about 2 per cent), so even a doubling of the protein content would make only a minute contribution to India's malnutrition problem. He claims that the effort to develop the protato was aimed more at gaining public acceptance of genetic engineering than at addressing the problem of malnutrition:

- "The cause of hunger isn't lack of food. It's lack of cash and of access to the food. Creating these GM crops is something to make them look attractive when actually the utility of eating them is very, very low. It's very difficult to see how this on its own will change the face of poverty".

Breeding and Reproducing Livestock and Fish

Biotechnology has long been a source of innovation in livestock and aquaculture production and processing and has had a profound impact on both sectors. Rapid advances in molecular biology and further developments in reproductive biology provide new and powerful tools for further innovation. Technologies such as genomics and molecular markers are valuable in understanding, characterizing and managing genetic resources in livestock and fisheries as well as in crops and forestry. Genetic engineering is also relevant in livestock and fisheries, although the techniques differ, and additional reproductive technologies are available in these sectors. This section describes the reproductive biotechnologies that are specific to the livestock and fisheries sectors.

The main objective of reproductive biotechnologies for livestock is to increase reproductive efficiency and rates of animal genetic improvement. The genetic improvement of locally adapted breeds will be important in realizing sustainable production systems within the broad spectrum of developing country production environments, and will probably best be realised by the strategic use of both non-genetic and genetic interventions. Reproductive biotechnology in fisheries presents opportunities to increase growth rates and improve the management of farmed species and to limit the reproductive potential of genetically engineered species.

As part of this country-driven strategy for the management of farm animal genetic resources, FAO invited 188 countries to participate in preparing the First Report on the State of the World's Animal Genetic Resources, to be completed before 2006. To date, 145 countries have agreed to submit country reports and 30 country reports have been received and analysed.

It is clear from these reports that artificial insemination (AI) is the most common biotechnology used by developing countries in the livestock sector. Many countries have requested training for the expansion of AI use, while expressing concerns that it has often been introduced without proper planning and may pose a potential threat to the conservation of local breeds.

Although the use of multiple ovulation followed by embryo transfer (MOET) is mentioned and the desire for its introduction or expansion expressed, no clear objectives for this technique are mentioned. All countries have expressed the wish to introduce and develop molecular techniques, often as a complement to phenotypic breed characterization.

Cryoconservation was identified as a priority by all countries and gene banks were recommended, but funding remains a major constraint. When animal GMOs are mentioned it is mainly to express the lack of proper regulations and guidelines for their eventual production, use and exchange. Some countries have expressed concerns that biotechnologies in the livestock sector should be, but are not always, pursued as an integral part of an overall genetic improvement strategy.

ARTIFICIAL INSEMINATION AND MULTIPLE OVULATION/EMBRYO TRANSFER

Advances in artificial insemination (AI) and multiple ovulation followed by embryo transfer (MOET) have already had a major impact on livestock improvement programmes in developed countries and many developing countries because they speed up the process of genetic improvement, reduce the risk of disease transmission and expand the number of animals that can be bred from a superior parent - the male in the case of AI and the female in the case of MOET. They also increase the incentives for private research in animal breeding and significantly expand the market for improved parent stock.

The number of AIs performed globally during 1998 was over 100 million in cattle (primarily dairy cattle, including buffalo), 40 million in pigs, 3.3 million in sheep and 0.5 million in goats. These figures illustrate both the higher economic returns in dairy cattle and the fact that cattle semen is much easier to deep-freeze than semen from other animals. Although over 60 million cattle AIs were performed in South and Southeast Asia, fewer than 1 million were performed in Africa.

AI is only effective when the farm sector has access to considerably greater technical and institutional and other organizational capacity than when male animals are used directly for breeding purposes. On the positive side, farmers employing AI do not have to face the costs or hazards of rearing breeding males and can have access to semen from any part of the world.

Despite the widespread use of AI in developed countries and in many developing countries, including within more advanced smallholder systems, it is applied only on farms that practise intensive livestock management with high-value animals. This is clearly not because of technical problems with semen production and storage, as most procedures are now fully standardized and proven to be effective under tropical developing-country conditions. Rather, it is because of the many organizational, logistical and farmer-training constraints that influence the quality and efficiency of the technology.

MOET takes AI one step further, both in terms of genetic gains possible and level of technical capacity and organization required. MOET is one of the basic technologies for the application of more advanced reproductive biotechnologies such as cloning and transgenics. During 2001 the number of embryos transferred globally was 450 000, mainly in dairy cattle, with North America and Europe accounting for 62 per cent, followed by South America (16 per cent) and Asia (11 per cent). About 80 per cent of the bulls used in AI are derived from MOET. The main potential advantage of MOET for developing countries will be in the possibility of importing frozen embryos instead of live animals, for example in the establishment of nucleus breeding stocks of locally adapted genetic resources, with the related lower sanitary risks.

Chromosome-set Manipulation and Sex Reversal in Fish

Controlling the sex and reproductive capacity of fish can be important for commercial and environmental reasons. One sex is often more desirable than the other; for example, only female sturgeon produce caviar and male tilapia grow faster than females. Sterility may be desirable when reproduction affects the taste of the product (*e.g.* oysters) or when farmed species (transgenic or not) might breed with wild populations. Chromosome-set manipulation and sex reversal are well-established techniques to control these factors. In chromosome-set manipulation, temperature, chemical and pressure shocks applied to fish eggs can be used to produce individuals that have three sets of chromosomes rather than the usual two. These triploid organisms generally do not channel energy into reproduction and thus are functionally sterile. Sex reversal can be accomplished by several methods including administering appropriate hormones. For example, genetically male tilapia can be turned into females through oestrogen treatments. These genetic males, when mated with normal males, produce a group of all-male tilapia.

Genetic Engineering in Livestock and Fish

Genetic engineering in animals can be used to introduce foreign genes into the animal genome or, alternatively, to "knock-out" selected genes. The method most used at present involves direct microinjection of DNA into the pronuclei of fertilized eggs, but progress is being made with new approaches such as nuclear transfer and the use of lentiviruses as DNA vectors.

In the first genetic engineering experiments with farm animals, genes responsible for growth were introduced into pigs to increase growth and improve carcass quality. Current research efforts include engineering resistance to animal diseases, such as Marek's disease in poultry, scrapie in sheep and mastitis in cattle, and diseases that affect human health such as *Salmonella* in poultry. Other examples include increasing the casein content of milk and inducing the production of pharmaceutical or industrial chemicals in the milk or semen of animals. Although conceptually simple, the methods used to genetically engineer livestock require special equipment and considerable dexterity, and no agricultural applications have proved commercially successful thus far. Applications in the near future therefore seem to be limited to the production of transgenic animals for use in the production of industrial or pharmaceutical products.

Genetic engineering is an active area of research and development in aquaculture. The large size and hardy nature of many fish eggs allow them to be manipulated easily and facilitate gene transfer by direct injection of a foreign gene or by electroporation, in which an electric field assists gene transfer. Gene transfer in fish has usually involved genes that produce growth hormone and has been shown to increase growth rates dramatically in carp, salmon, tilapia and other species.

In addition, a gene from the winter flounder that produces an antifreeze protein was put into salmon in the hope of extending the farming range of the fish. The gene did not produce enough of the protein to extend the salmon's range into colder waters, but it did allow the salmon to continue growing during cold months when non-transgenic salmon would not grow. These applications are still in the research and development stage, and no transgenic aquatic animals are currently available to the consumer.

SOCIAL RESISTANCE TO AGRICULTURAL BIOTECHNOLOGY

Over the last two decades, the life sciences industry has made enormous investments in biotechnology research and development; thrown tremendous energy into getting its genetically engineered (GE) crops approved, patented, and commercialized; and lobbied U. S. farmers and food producers to use them. For their part, the U. S., British, and other governments that envision the biotechnology sector as the wave of the future, and as a means of augmenting their national competitiveness, have strongly supported the industry and its efforts to commercialize (and normalize) these new technologies.

They have devoted considerable sums of money to biotechnology research, taken significant strides to deregulate the industry, and sought to promote the spread of U. S.-style intellectual property rights in the World Trade Organization. The U. S. government in particular has also promoted the dissemination of agricultural biotechnology in developing countries through the U. S. Agency for International Development.

With so much economic and political muscle propelling them, it is not surprising that GE crops hit the ground running when they came onto the scene in the mid-1990s. But what is surprising is that the rapid growth in GE crop deployment has been matched by an equally remarkable (and perhaps historically unprecedented) proliferation of citizens' voices challenging the biotechnology industry on economic, environmental, cultural, and moral grounds. Indeed, long before transgenic crops made their way to the market, individuals and groups concerned about the dissemination of these new technologies were already questioning their safety, utility, and necessity.

In this chapter, we explore the character and impact of the new social movement against genetic engineering in agriculture (hereafter, the "antibiotech movement"), as it is unfolding in the North, which is the heart of the agricultural biotechnology industry and the main arena in which these technologies have thus far been deployed. We analyse the strategies social activists have employed in efforts to limit the deployment of biotechnology in food and agricultural production, and their effectiveness in achieving this goal.

This antibiotechnology activism, we contend, has turned what until recently looked like a done deal in the trajectory of agricultural industrialization—the shift from farming to biotechnology, as David Goodman and his colleagues have put it — into a moment of uncertainty and openness. Activism has created this uncertainty, we argue, by affecting the fortunes of the biotechnology industry and the political environment in which it is developing in three substantial ways.

First, activism has forced states to reconsider their liberal regulatory approaches to agricultural biotechnology, and to take steps to regulate the industry more seriously. Although pushing recalcitrant states toward stronger regulatory oversight remains difficult, there are signs that morestringent national-level regulation will be forthcoming in many countries, including the United States. Second, antibiotech activism is helping to spawn new global regulatory regimes for genetically engineered organisms, such as the Cartagena Protocol on Biosafety and new Codex Alimentarius Commission standards on bioengineered foods.

Although social activists are not the main actors negotiating these regimes, they have helped ensure that their content moves the world toward more substantive regulation of these organisms. Third, the antibiotechnology movement has imposed important direct and indirect costs on the firms developing this technology. In addition, organized social activism has moved the issue of agricultural biotechnology out of relative obscurity, and out of the hands of a small number of corporate and state actors, into the public arena, where it is being debated by a broader spectrum of society.

In making this argument, we want to emphasize two points with respect to the recent sociological (as well as popular) literature on the industrialization of agriculture, of which biotechnology can be seen as the latest manifestation. First, in contrast to those who suggest that the events of the past several decades have provided transnational corporations with virtually unchallenged power to determine the future of world agriculture, we argue that social activists and their organizations are critical actors that have been overlooked by analysts of world-food-system restructuring.

Without exaggerating their power with respect to large corporations and powerful states, we argue that social activists are playing an underappreciated role in shaping patterns of food production, consumption, and distribution. In the case considered here, antibiotechnology activists have placed new demands on biotechnology firms and, even more significantly, on states, demands that have forced them to abandon certain paths and move in new directions.

Second, we want to distinguish ourselves from those who see consumer resistance as the primary basis for the recent change in the political and economic climate surrounding agricultural biotechnology. Although we do not mean to ignore the influence of consumer opposition, we hold that it is social-movement mobilization that has made consumers aware of genetic engineering of the food supply and that has motivated consumers to take a stance. Even in Europe, where consumer opposition has been energized by food and health scares such as mad cow disease in Britain, AIDS-tainted blood in France, and a dioxin scare in Belgium, it is activist nongovernmental organizations (NGOs) that have made agricultural genetic engineering into a social issue. We begin by providing a sketch of what the antibiotech movement looks like and who it comprises.

We then discuss the three main arenas in which the movement has concentrated its efforts in the North—the domestic regulatory sphere, the international regulatory sphere, and the corporate sphere—and evaluate its effectiveness. Finally, we consider the strengths and limitations of this type of activism.

FLUTTERING BENEATH THE MONARCH SUITS

In the mainstream U. S. media, one is most likely to encounter three rather different representations of antibiotechnology activism. One projects an image of playful public protest, such as the piece of street theater outside the Food and Drug Administration's (FDA) 1999 public GE hearing in Chicago, which featured a human fish-tomato, a grotesque cow that was shot up with genetically engineered bovine growth hormone by a mad scientist, and eight children dressed as monarch butterflies who were symbolically killed by genetically engineered mutant corn pollen.

This sort of action has created a distinctive iconography for the antibiotechnology movement, in which the monarch butterfly has become a prominent symbol. Thus, in the popular street politics that has become commonplace in Northern countries since the 1999 WTO protest in Seattle, the antibiotechnology movement has a discrete recognizable presence that is not reducible to the antiglobalization sensibility into which many observers lump this politics.

In fact, many movement participants are associated with local foodsafety organizations or are organized in local or regional activist networks, such as the U. S. RAGE (Resistance against Genetic Engineering) and British GenetiX Snowball networks. They mobilize for specific campaigns and have played a major role in grassroots civic and protest activities such as "raiding" supermarkets and labeling GE products.

The second representation is less forgiving. It depicts "special interest terrorists" and "eco-warriors" who clandestinely try to derail the biotechnology industry by vandalizing research facilities and destroying seed trials. Such actions are generally undertaken by local grassroots groups (some with rudimentary organizational bases) that often sport rebellious names, such as the Dusty Desperadoes and Strawberry Liberation Front in northern California, and GrainRage and the "elves" of the Earth Liberation Front in Minnesota.

By some accounts these groups represent no more than an alternative youth culture with anarchist leanings. They have fueled an image of nocturnal sabotage, illegality, and antisocial mayhem that has led legislatures in several

U. S. states to press for tougher law enforcement penalties in property-destruction cases. Indeed, in February 2002, the Federal Bureau of Investigation singled out the Earth Liberation Front as a particularly dangerous domestic organization, one that poses a serious terrorist threat to the United States. They are, in fact, a fringe element in the movement, though they have been expensive to the industry and, in some people's opinions, to the movement as well.

The third image is one of antiscience, antiprogress Luddites, an image that to some observers also connotes a mysticist sensibility. This image is frequently invoked, especially by scientists, who express anger at the critics of biotechnology. A good example is Norman Borlaug's characterization of the "antiscience zealotry" that drives antibiotech activists.

To many scientists, the extraordinary risk-aversion of biotech's critics and their insistence that we should step back from the potentials of scientific discovery simply because they fear there is a genie in the bottle are irrational, nonor antimodern, and selfish. Yet those who frame the issue in this way fail to recognize that a significant segment of the antibiotech movement also comprises intellectuals, researchers, social scientists, and natural scientists working in established institutional settings on a range of issues.

Many antibiotech activists are scientific experts in their own right, such as Rebecca Goldburg at Environmental Defence, Jane Rissler and Margaret Mellon at the Union of Concerned Scientists, and Michael Hansen from Consumer Policy Institute. The movement also maintains connections with university scientists. The importance of this small but crucial group of activist-experts is that they are able to engage in the politics of expertise, framing their claims to authoritative knowledge in the terms used by industry scientists but presenting the issues from the perspective of the public interest.

All these images of antibiotech activism, taken separately or together, are at best partial, and even misleading. What the images do indicate is that the movement is, both strategically and organizationally, diverse and complex. Strategically, the movement works in a rather loosely connected bricolage pattern of action: there are campaigns, labeling drives, protests, trial-crop destructions, and also serious research and knowledge-generation efforts to counter science with science.

Organizationally, the movement comprises an extensive web of activist networks that engage a wide array of civil-society organizations connected not only across the world but also across a variety of issues, social concerns, and interest groups. This network mobilizes the resources of both mainstream and radical environmental NGOs, running the gamut from the Sierra Club and the British Soil Association to Greenpeace and Friends of the Earth International.

It includes science based groups, such as the Council for Responsible Genetics in the United States; the Research Foundation for Science, Technology and Ecology in India; and the Institute for Science in Society in the United Kingdom. It links the lobbying capacity of food policy NGOs such as the Institute for Development Policy ("Food First") and the Institute for Agriculture and Trade Policy in the United States with Southern activist groups such as the Tamil Nadu Women's Forum in India, Bio Watch South Africa, and MASIPAG in the Philippines.

It engages farmers' organizations in the North as well as the South, such as the Confederation Paysanne in France and the New Agriculture Movement in Bangladesh. In short, the network nodes of antibiotechnology mobilization extend across local, national, regional, and transnational organizational structures.

In a movement of this kind, organizational leadership and unity give way to coordination and decentralization. This means that local-level activism has cultural and political dynamics that go beyond street theater, mass protest, and the illegal destruction of trial plots. In Britain, the United States, and Canada, activists have been arrested inside supermarkets for affixing warning labels to products containing GE ingredients. In Ecuador and Brazil, activist groups have prevented ships carrying GE grain from docking. In the United States, local lobbying campaigns have persuaded city councils in Austin, Boston, Boulder, Ann Arbor, and other places to require GE food labeling.

Moreover, local actions are hooked into a larger panoply of activities facilitated by the network structure. In particular, activist organizations have used communication networks and the Internet to advertise protests and other actions, to generate petitions, and to solicit letters of protest against governments and firms around the world. Internet sites such as Resistance is Fertile and GenetiX Alert were established exclusively as clearinghouses of news and information about antibiotechnology actions.

Such informational networks have facilitated creative collaborations between NGOs and local activists across the North and South. The United States-based Institute for Agriculture and Trade Policy, for instance, supported a long march for sustainable agriculture in Thailand; and Greenpeace has been a ubiquitous campaigner, assisting local actions from Boston to Bombay. In addition, these networks have enhanced the ability of NGOs and advocacy groups in the North to coalesce and share resources. In July 2000, for instance, seven major public-interest NGOs in the United States formed the Genetically Engineered Food Alert to press for a moratorium on genetically engineered foods until they are safety-tested and labeled and until biotechnology companies accept liability for any harm to human health or the environment.

The coalition's strength is its ability to coordinate the resources and skills of the different organizations while moving information rapidly between them in order to build public support and sustain pressure on the biotechnology industry. As with any social movement, the prospects of the antibiotech movement depend on its ability to maintain its momentum. Although the future is hazy, activists have become a real force, and this complex, multifaceted movement has real institutional strengths.

These strengths are manifested particularly in the ability of organizations to coalesce and to apply the weight of their resources and capacities—finance, human power, professional skills and training, infrastructure—synergistically to the pressure points of regulation and profitability in the deployment of agricultural genetically engineered organisms (GEOs). In so doing, these organizations engage two core principles of a democratic polity: accountability and choice.

ARENAS OF ACTION AND ACTIVISM

In trying to limit the production and trade of GEOs, antibiotechnology activists have focused their efforts in three broad arenas. One is the domestic regulatory arena, where activists are pressuring the state to become a more serious regulator of biotechnology; the second is the international arena, where activists are working to create new supranational regulatory regimes for the global deployment of GEOs; and the third is the economic arena, where activists have sought to pressure biotechnology firms and consumers of their products to change their behaviour. In each of these arenas, antibiotechnology activists have developed carefully thought-out strategies and achieved significant successes.

Holding the State Accountable

Although some of the globalization literature claims that states have been rendered powerless by globalization, this is clearly not true with respect to the regulation of biotechnology. States continue to exert significant leverage over the biotechnology industry and its behaviour at the national level, particularly in the advanced industrialized countries where biotechnology has been most aggressively developed and deployed.

It is true that governments in some advanced capitalist countries—especially the United States, Canada, and Britain—have chosen to regulate the agricultural biotechnology industry only minimally and to encourage its growth. In this context, antibiotech activism has concentrated on exposing the dangers inherent in such a liberal regulatory approach and on pushing governments to adopt more comprehensive and responsible regulatory policies.

On both sides of the Atlantic, activists have revealed the weaknesses of existing regulatory structures by making embarrassing public revelations of government practices. In May 2000, for instance, activists in the United Kingdom seized on the fact that a Canadian company had inadvertently (and illegally) sold conventional rapeseed (canola) contaminated with Monsanto's Roundup Ready rapeseed to British farmers, who unknowingly planted it on more than ten thousand acres. Activists used the incident to claim that the technology is effectively uncontrollable and that the government was overly complacent in providing no oversight. (The problem had, in fact, been discovered not in Britain but in German field tests.)

In an effort to minimize the seriousness of this event, government politicians initially insisted that the contaminated seed was effectively sterile and unlikely to cross-pollinate. But activists' ability to expose the government's low level of knowledge, its weak regulation, and its tendency to take industry claims at face value severely undermined its popular credibility.

Just four months later, antibiotech organizations in the United States made similar hay of the StarLink taco debacle, described in the introduction to this volume. In this case, activists pioneered the use of a new tactic: paying for independent testing of food products to see whether they contain genetically engineered ingredients. Indeed, it was the antibiotechnology coalition Genetic Engineering Food Alert that hired an Iowa-based genetics-testing firm, headed by a sympathetic scientist, to test whether several commercial brands of tacos shells contained StarLink.

After determining that the taco shells did indeed contain the unapproved corn, the coalition shared this information with the press and took the U. S. government to task for creating a regulatory system that could allow a bioengineered product to be approved for one type of use (animal consumption) but not another (human consumption), with no effective means of keeping the two separate. In addition to the tremendous economic costs that the StarLink debacle imposed on the industry, retailers, farmers, and grain traders, the incident pushed the issue of agricultural biotechnology—and the weakness of government oversight—squarely into the public eye.

In both the rapeseed and taco cases, antibiotechnology activists stressed the difficulty of controlling this technology (especially under a free-market system) and highlighted the responsibility of the state to do so. They have complemented this strategy with sustained litigation. Court action is a powerful instrument for raising the stakes of public accountability in the foodregulation regime because it enables activists to frame specific and sometimes technical questions in terms of the state's responsibility to the public.

For instance, in 1999, a coalition of activist organizations and citizens filed suit against the FDA, seeking to overturn its 1992 policy of neither labeling GE foods nor subjecting them to special testing. The litigants claimed that the FDA's decision violated the Federal Food, Drug, and Cosmetic Act, the agency's principal regulatory instrument. The point of the lawsuit was to show that the FDA had failed to ensure the safety of consumers regarding GE foods because the available scientific knowledge did not sustain the assumption of "substantial equivalence" and because safety tests carried out by the industry were tailored to meet government criteria that were designed to favour the industry. Thus, the suit was not about the safety of GE foods per se, but about the adequacy and accountability of government regulatory procedures. One consequence of the suit was that it forced the government to place in the public domain a vast amount of information about internal disagreements, decision-making procedures, and the bases of scientific findings, information that compromised the confident tone of its regulatory pronouncements.

Activists have also pressed for public accountability with respect to the environmental impact of the new agricultural biotechnologies. In 1997, seventy nonprofit organizations filed a petition against the U. S. Environmental Protection Agency (EPA) in an effort to stop the release of crops modified with the bacterium *Bacillus thuringiensis*, or Bt crops. Supporters of the technology argue that Bt is a boon to the environment because its built-in resistance to pests enables farmers to use much less insecticide.

Opponents argue that there is a good chance that pests will become tolerant to the Bt toxin, thereby reducing its effectiveness and its utility in organic farming. An additional concern is the potentially harmful effects on other species, such as the monarch butterfly larvae, that were indicated by a study at Cornell university. The petition called on the EPA to withdraw all current registrations and deny future approvals of Bt crops, pending appropriate evaluation of environmental risks and economic impacts on organic growers.

This petition was unsuccessful. But it demonstrates how such actions can increase public pressure on government agencies to exercise stronger oversight. As part of the petition, Greenpeace commissioned a review of the impact of transgenic Bt plants on nontarget organisms. The review, carried out by an independent group of Swiss scientists, was highly critical of the industry's scientific tests, which provided the basis for the EPA's approval for insect-resistant crops.

Though the EPA continues to insist that its regulatory process ensures that these products are safe for human health and the environment (partly through a controversial and costly "refuge" requirement), the agency was forced onto the defensive and agreed to undertake a comprehensive review of its policies in 2000–2001. The FDA and the U. S. Department of Agriculture (USDA) also reviewed their agencies' regulatory procedures during this same period, largely in response to activist scrutiny and pressure.

In tandem with these "risk and procedure" cases, activists have also pushed "risk and rights" cases against state regulatory agencies. In particular, activists have argued that clear labeling of GE and non-GE foods would allow consumers to make informed choices about their own level of acceptable risk. Activists have pushed hard on this front because the political logic is inexorable: that citizens should have the right to know what they are ingesting is so obvious that almost any argument for curtailment advanced by the industry or by the government seems suspicious.

Indeed, repeated polls have indicated a widespread public preference for labeling. In March 2000, the Centre for Food Safety (CFS) spearheaded a large coalition of organizations that filed a legal petition with the FDA demanding the development of a thorough premarket environmental-testing regime and mandatory labeling for GE foods. The CFS built its argument on consumers' right to know which foods have been genetically engineered and their right to make an informed choice.

This tactic aimed to inject citizens' rights into the struggle over regulation as part of a broader strategy to force government agencies to articulate their responsibilities to citizens in the authoritative arena of the court and to pressure retailers to treat their consumers with respect.

How effective have such actions been? In Europe, activist-inspired public pressure virtually forced the fifteen governments of the European Union to stop approving new GE crops in 1998—a de facto moratorium that effectively persists today. And in the United States, some fifty thousand public comments to the FDA (many of them the result of organized petition drives) and three heated public meetings in 1999 pushed that agency to develop a more stringent policy proposal for approving genetically engineered plants.

In 2000, then secretary of agriculture Dan Glickman initiated a review of the USDA's GE-related policies and procedures by the National Research Council of the National Academy of Sciences. The review was critical of current regulatory procedures, specifically calling for greater public participation and interdepartmental scientific review in the regulatory process. Though it is still too early to discern the actual impact of the independent review, it might lead to some policy changes.

Given the power and authority of the organizations involved in the development of biotechnology, these shifts are important markers of political struggles over the deployment of GEOs, especially in the United States, where industry strategists have relied on high levels of public inertia. To be sure, part of the motivation for reconsidering regulation has been the rising outcry against GEOs in Europe, the threats to trade, and the falling sales faced by biotechnology companies.

But these developments have been greatly advanced by the actions of public advocacy and environmental activist groups. In effect, activist initiatives have forced government regulatory agencies to respond to their efforts to shape a critical public consciousness of biotechnology's social impact and their efforts to force the state into a tighter regulatory role.

Regulating beyond the Nation-State

Activists have also sought to slow the biotechnology train by helping to create supranational regulatory regimes that govern the international deployment of GEOs in food and agriculture production. This strategy is reflected in the movement's efforts to create an international protocol (known as the Cartagena Protocol on Biosafety) regulating trade in transgenic organisms, as well as in its ongoing efforts to influence the work of the Codex Alimentarius Commission, the UN organization that sets international standards for food health and safety.

The idea for an international biosafety protocol was born during the negotiation of the UN Convention on Biological Diversity in 1992. During these negotiations, participants from the global South pointed out that their rich stores of biological diversity were uniquely vulnerable to the introduction of genetically engineered organisms. For this reason, Southern participants saw an urgent need for a protocol to oversee international trade in GEOs, which threatened to increase dramatically in the coming years.

But the United States and a handful of its allies (known collectively as the Miami Group) perceived such a protocol as a threat to the "right" of Northern farmers and corporations to produce and sell their goods, including genetically engineered agricultural commodities, worldwide and without prejudice. The group thus did its best to prevent the negotiation of such an agreement.

Ultimately, the powerful Miami Group lost. Although the final agreement was not as strong or comprehensive as most of the South wanted, it was broadly perceived by Southern negotiators and the antibiotech movement as a victory, since both groups had been seeking at least some form of control over GEO trade. Moreover, the fact that the regulatory regime would operate under the auspices of the relatively inclusive and democratic UN was perceived as a significant achievement, as the Miami Group had made a preemptive move to shift the onus of regulation to the clearly undemocratic World Trade Organization (WTO).

Although many factors contributed to the successful negotiation of the Cartagena Protocol in February 2000, the role of social activists was crucial. Prior to the negotiations, European activists and publics had put tremendous pressure on their governments to take genetic engineering seriously and to adhere to the "precautionary principle". This pressure induced the European Union to dispatch a number of high-level ministers to the negotiations, sending a clear signal to other parties (including the United States) that Europe intended to be responsive to its citizens.

Representatives from a wide range of NGOs, from the North and South, also attended the meetings in person and used a variety of tactics to pressure delegates to reach a meaningful agreement. The NGOs engaged in a division of labour, sending some members into the meetings as formal observers and spokespersons and others onto the streets to demonstrate. Activist groups organized all-night vigils in the freezing Montreal weather, trotting out one colourful image after another to keep the media's attention.

Activists on the inside also worked around the clock, lobbying delegates and disseminating up-to-the-minute economic, legal, and scientific information. Shortly after the negotiations began, the activist network released a position statement, signed by twenty-five influential NGOs from around the world, that demanded a strong biosafety protocol and admonished the Miami Group for obstructing the negotiations. Activist organizations also sought to shore up the power of Southern delegations. Once it became clear that the United States had no interest in listening to the activists' position, they focused their efforts on forming a transnational alliance with state actors from the main negotiating bloc for the South, known as the LikeMinded Group. The activists supplied these Southern delegates with information, helped them assess proposed text, and made concrete recommendations about accepting, rejecting, or changing specific language. They also brought environmental scientists to the meetings to give public presentations, and invited the delegates to attend.

Reflecting on the negotiations, Tewolde Egziabher, the chief negotiator from the South, highlighted the importance of the activists' support when he explained how the Like-Minded Group, composed largely of African countries, managed to achieve anything against foes as powerful as the United States and its allies. "We had friends," Egziabher wrote.

Africa is financially so poor that the African Group would not have functioned ... without friends. But we soon made friends who filled in our gaps.... the Third World Network gave us critically needed assistance, and facilitated critically needed interactions, both South-South and South-North. African telecommunications are so poor that had it not been for the Gaia Foundation of London acting as an information relay station, we could not have been effective. And had our many, many other friends all over the world not helped, we would not have managed to stay as informed and as effective as we did.

As one observer put it, the activists "were the eyes watching for society at large, calling delegates on the carpet if they didn't remember that they have citizens back home who might not be happy if they do certain things.... They were the watchdogs."

Many of the same organizations are playing a similar watchdog role in the context of the Codex Alimentarius Commission. In 1999, the Ad Hoc Intergovernmental Task Force on Foods Derived from Biotechnology was set up within the Codex to develop standards and guidelines for foods derived from biotechnology.

The definition of these standards and guidelines—which include risk-assessment criteria and labeling requirements and which recognize the role of the precautionary principle in food safety standard setting—lies at the heart of the political struggles over the task force's work. Activists have followed the committee's deliberations closely, studying and critiquing the official positions taken by governments and their representatives and pressuring governments and the task force itself to try to influence its decisions. A central goal of these activist groups is to ensure that task force decisions do not contradict or override the biosafety protocol's provisions.

Activists are of course not the only ones trying to affect these decisions; every group with a strong interest in the Codex's decisions has sought to get its voice heard. In fact, one could argue that the activists' efforts are simply aimed at counteracting the influence of far more powerful Northern agribusiness interests.

Generally speaking, however, activists' efforts to establish and shape new international regulatory regimes for genetically engineered foodstuffs have helped to make decision making about agricultural biotechnology more democratic by ensuring that decisions about the deployment of biotechnology are not left entirely up to those with the most to gain from promoting these products and the least to lose if the products have negative economic or environmental consequences in importing nations.

Putting the Industry on the Defensive

Social activism has not only forced the issue of biotechnology firmly on to the regulatory agenda but also complicated the economic lives of the biotechnology corporations. In recent years, church groups, consumers, farmers and farmers' associations, chefs, food processors, food retailers, and professional environmental and food-policy activists have all put direct pressure on these corporations, forcing them to rethink their strategies and, in some cases, to change their behaviour.

Perhaps most striking about the experience of the biotechnology industry in the 1990s is how unprepared it was for the strong reaction to its new attempts to "improve" nature. As one economist dryly observed, "Monsanto didn't even know a train was *coming* before it got run over." This may explain in part why the industry was so slow to react to the barrage of criticisms, consumer campaigns, lawsuits, and other actions that have thrown this techno-scientific project into question. But another part of the explanation surely resides in the divergent character—and preparation—of the two adversaries in this intense struggle for public opinion. On one side is a group of scrappy and diverse activist organizations who can punch and jab faster than you can say "genetic engineering," and on the other side is an elephant tethered to a post by its size, its enormous and sunk costs, and its competitive character.

Moreover, activist organizations have spent the last twenty years in basic training for just such a battle, one that requires particular political, organizing, and issue-framing skills and occurs on many different fields simultaneously. By contrast, many life sciences corporations were so preoccupied with building up their strategic positions in this long lead-time, capital-intensive, and risky industry and with trying to control the regulatory agenda that they barely knew they had an adversary in the making.

From this position of strategic if not economic advantage, antibiotechnology activists and other concerned groups have launched an all-out offensive against firms situated at virtually every point along the commodity chain. They have instigated shareholder actions against publicly owned corporations, demanding that they disavow the use of genetically engineered inputs.

Activists have also exerted pressure through lawsuits, such as one filed in 1999 by several prominent antitrust lawyers against Monsanto on behalf of small farmers and farm groups. The suit, inspired by a coalition of environmental groups, accused the biotechnology giant of being at the centre of an international conspiracy to control a large part of the world's seed supply and of giving farmers "false and fraudulent guarantees about the safety and marketability of a new breed of bioengineered seeds".

Activists have also organized consumer campaigns to convince food processors, retailers, and restaurant chains to stop selling genetically engineered foods, under the assumption that the market signals would quickly travel back up the supply chain to exporters, farmers, and, ultimately, the life sciences companies. In response to consumer activism, a number of major supermarket chains, particularly in Europe, started to remove GE products from their shelves as early as 1998, and some North American and Canadian chains are now following suit.

The Organic Consumers Association and the BioDemocracy Campaign have also targeted fifteen food companies and distributors—dubbed the "Frankenfoods Fifteen"— to get them to stop using GE crops in their products. In general, these campaigns have been surprisingly successful. Since 1998, some of the world's major food processors, including H. J. Heinz, Gerber, Bestfoods, Frito-Lay, Unilever, Seagram's, and Nestle, have publicly declared their products to be GE-free in some or all their markets. And in August 2000, a major biotechnology producer, Novartis, declared that it was eliminating GE ingredients from all of its consumer food products—in effect refusing to buy its own products.

Consumer pressure—or the fear of it—has affected the whole food system. In 1999, McDonald's, fearing a threatened boycott, decided not to use genetically engineered potatoes in its French fries. After similar decisions were taken by Wendy's and Burger King, three major potato processors— J. R. Simplot, McCain Foods, and Lamb-Weston—advised farmers they would not be buying Monsanto's genetically engineered NewLeaf potatoes the following year. In consequence, Monsanto closed the research lab in Maine where the NewLeaf was developed and ultimately, pulled the potato from the market.

In Europe, these campaigns have succeeded in part because they tap into a smoldering anger about U. S. food imperialism. This anger was graphically depicted in 1999 when activist farmers dumped four tons of GE grain in downtown London from a truck bearing a large sign that read "Tony, Don't Swallow Bill [Clinton]'s Seed. " When the French farmer José Bové, who trashed several McDonald's restaurants in his country and became a kind of folk hero, was asked what motivated his extreme actions, he replied, "To fight against globalization and advance the right of people to eat as they see fit". In the United States, activists have gathered support for their corporate campaigns by appealing to people's health concerns.

One poster printed up by Greenpeace depicts a Kellogg's cereal box, with a caption reading "Kellogg's Frosted Fakes." The Kellogg's mascot Tony the Tiger also appears, saying "Frankenfood BAD!" In general, activists have portrayed the industry as greedy and irresponsible, more concerned with profits than with public health, the environment, or poor countries' food security.

U. S. markets in Europe, Japan, and other countries have been significantly affected by the negative reaction to GEOs in the food supply. Between crop years 1997–98 and 1998–99, for instance, European imports of U. S. corn fell precipitously, from 2 million to 137,000 tons. Worried about losing their export markets, some farmers are adopting a more cautious attitude toward new GE crop varieties, such as GE wheat. Gary Goldberg, CEO of the American Corn Growers Association, argues that farmers' reactions are motivated by their economic concerns: "For agricultural producers, this debate over GEOs is not a safety, environmental or health issue. It is an economic issue.... Simply put, can farmers afford to grow a crop that they may not have a market for?"

These pressures have made it difficult for biotechnology firms to proceed with business as usual. Evidence that the biotechnology industry has been hurt is apparent both in the defensive moves the big firms have made in the last few years and in the industry's all-out effort to redeem itself in the eyes of the public, its direct customers (farmers), and its investors. For years, the industry kept a low profile, quietiy developing and commercializing its new gene technologies, but that strategy has become impossible. The industry effectively admitted as much in November 1999, when seven of the largest life sciences firms formed a new industry alliance, the Council for Biotechnology Information (CBI), to counter criticism from antibiotechnology organizations. "The protest industry has gone too far," Edward Shonsey, a chief executive at Novartis Seed, told a *New York Times* reporter in November 1999. "They've crossed the boundaries of reasonableness, and now it's up to us to protect and defend biotechnology". Through the CBI, the industry launched a $50 million advertising campaign to win back control of the discourse about agricultural biotechnology. Running the ad campaign are several high-powered public relations firms.

Individually and collectively, the life sciences firms are seeking to depict themselves as firmly committed to improving the social condition. One DuPont television commercial identifies one of its future tasks as "finding food to fight breast cancer. " Another portrays a poor old woman carrying a cooking pot in an unnamed Southern country; the caption proclaims that biotechnology will be able to "make food grow where food can't".

The health and nutrition page of the CBI's Web site highlights four areas in which the industry is poised to improve the human condition: by producing foods that will deliver oral vaccines, reduce or eliminate allergies, be more nutritious, and last longer and taste better. In addition, the life sciences firms are undertaking highly publicized goodwill efforts in the battle to remake their image. In early April 2000, Monsanto announced that it had a complete sequencing of the first crop-plant genome (rice) and then made this information publicly available by donating it to the International Rice Genome Sequencing Project, a consortium of ten publicly funded genome centres based in Japan. A month later, AstraZeneca announced that it would help make a genetically engineered strain of "golden rice" freely available to the developing world.

But perhaps the most significant evidence that social activism and consumer concerns are hurting the biotechnology industry financially is the shake-ups that have occurred in the biotechnology stock market and the fortunes of individual companies. Monsanto's stock price fell by half in 1999 because of the company's troubles in the agricultural biotechnology arena. In March 2000, Monsanto was compelled to merge with Pharmacia & Upjohn, which then hived off the agricultural part of its operation and put it on the market.

AGRICULTURAL BIOTECHNOLOGY REGULATION

For traditionally bred plants, regulators rely on plant breeders to conduct appropriate safety testing and to be the first line of defense against genetic alterations that might prove dangerous. The same should be true for agricultural products developed using biotechnology. There is no reason for regulators to treat biotech plants any differently than traditionally bred plants, particularly given the fact that biotechnology provides greater control over gene manipulation. Despite fears about gene manipulation, traditional crossbreeding has altered plant genes and improved the human diet for the past 30 years. In fact, people worldwide safely consume these plants every day. Examples include wheat, corn, potatoes, tomatoes, and countless other staples of the American diet. In 1984, the biotechnology industry began experimenting with "genespliced plants" — the more advanced approach to gene manipulation that we now call "biotechnology."

REGULATORY SCHEME

At the time, the White House Office of Science and Technology Policy began crafting a frame-work wherein existing federal agencies would regulate genetically engineered organisms on the basis of their characteristics, not the method of production — thus wisely deferring to the scientific community's judgment that regulation ought to address the products, not the process of biotechnology. Engineered organisms would not require extra scrutiny simply because genetic engineering produced them (the process). Instead, they would be subject to heightened scrutiny only if the individual organisms expressed characteristics that posed some conceptually heightened risk (the product). The federal government divided regulatory jurisdiction among agencies already involved in agricultural, food, and environmental regulation. These include the United States Department of Agriculture (USDA), the Environmental Protection Agency (EPA), and the Food and Drug Administration (FDA). While each of these agencies considers the characteristics of individual products in their regulation, only FDA followed the general scientific thinking that genetically engineered and non-genetically engineered products should be regulated similarly. Both USDA and EPA automatically subject all genetically engineered plants as a class to premarket approval requirements not ordinarily applied to conventionally bred plants.

DEPARTMENT OF AGRICULTURE

USDA regulates the release of all genetically engineered agricultural plants under statutes giving USDA's Animal and Plant Health Inspection Service (APHIS) the authority to regulate plants that may be or may become weeds or other nuisances — what the statutes call "plant pests." Although the rules apply in a general sense to novel or exotic varieties of both genespliced and conventional plants, APHIS typically only requires field testing of conventional plants that are new to a particular U.S. ecosystem (transplanted from another continent, for example).

However, all genetically engineered agricultural plants face a higher regulatory bar — making it more difficult to expand the use of biotechnology. Genetically engineered crops must be field tested under APHIS regulations prior to commercialization. For example, a new variety of corn produced with conventional hybridization requires no government-mandated field testing, but all new varieties of genetically engineered corn do, even though there is no logical reason for the regulatory disparity. For most genetically engineered plants, APHIS requires the company producing the plants to submit notice detailing the gene or genes that have been inserted, where the plant will be tested, and other relevant characteristics of the plant prior to receiving permission to conduct the field trials. Once the company completes field testing, APHIS reviews the results and makes a determination on whether or not the product should be "deregulated" and can be released into the market.

ENVIRONMENTAL PROTECTION AGENCY

EPA regulates plants that are modified to produce substances that act like pesticides: that is, substances used by a plant to protect itself from pests, such as insects, viruses, and fungi. EPA's proposed rule for these "plant-incorporated protectants" is not yet finalized. In the interim, however, EPA has regulated such plants by applying its proposed guidelines, which are functionally similar to rules for the registration of synthetic chemical pesticides. Again, biotech products face higher regulatory hurdles, even though plant-incorporated protectants developed through conventional breeding are exempt from these requirements.

FOOD AND DRUG ADMINISTRATION

FDA is responsible for ensuring that food items, including foods derived from genetically modified plants, are safe to eat. Following the general regulatory framework that emphasizes product regulation rather than process regulation, FDA rightly does not treat foods derived from genetically engineered plants to be inherently unsafe. Food producers are not required to seek premarket approval from FDA unless there is a substantive reason to believe that the novel trait (or traits) in the food pose a safety question.

The initial determination of safety is left to the producer, but FDA has encouraged producers to consult with agency scientists prior to marketing a food produced with biotechnology to ensure that the appropriate determination is made. Recently, FDA published a proposed rule that would require producers to notify the agency at least 120 days before marketing, which may be a signal that FDA could abandon its more reasonable approach in the future. In addition, FDA does not require labeling of foods derived from biotechnology unless the genetic insertions so alter the food that the common name no longer describes it adequately. Examples of this phenomenon would include such alterations of the food product that would raise a safety issue or change the product's nutritional composition or its storage or preparation characteristics.

LABELING ISSUES

Some activists, however, argue that the government should mandate the labeling of all genetically engineered foods. They assert that consumers have a "right to know" how their foods have been altered, and that a mandatory label would best allow consumers to choose between genetically engineered and conventional foods.

Biotechnology advocates have argued against mandatory labeling because such requirements raise food costs — something that mostly harms lower-income Americans and people on fixed budgets. Perhaps more important, while biotech products are not substantially different from other products, special labels would likely make consumers think these products were more dangerous.

Hence, rather than serving educational or "right to know" purposes, such labels promise to simply confuse consumers.

A government-mandated label on all genetically engineered foods also would raise important First Amendment free speech issues. In 1996, the U.S. Court of Appeals, in *International Dairy Foods Association et al. v. Amestoy*, ruled unconstitutional a Vermont statute requiring the labeling of dairy products derived from cows treated with a genetically engineered growth hormone, noting that food labeling cannot be mandated simply because some people would like to have the information. "Absent ... some indication that this information bears on a reasonable concern for human health or safety or some other sufficiently substantial governmental concern, the manufacturers cannot be compelled to disclose it."

In other words, to be constitutional, labeling mandates must be based in science and confined to requiring disclosure of information that is relevant to health or nutrition. Furthermore, consumers need not rely on mandatory labeling of biotech foods to truly have a choice. Realworld examples show that market forces are fully capable of supplying information about process attributes (including kosher and organic production standards) that consumers truly demand. The same can be said about nonbiotech foods, and the FDA recently published proposed guidelines to assist producers in voluntarily labeling both genetically engineered and nongenetically engineered foods. Additionally, the USDA's newly published rule for organic certification necessarily excludes biotech products from organic food production.

Consequently, consumers wishing to purchase nonbiotech foods need only look for certified organic products.

INTERNATIONAL TRADE

While U.S. consumers do not appear to be strongly opposed to biotech foods (they, in fact, seem rather indifferent), a strong anti-biotechnology movement has arisen in several European countries over the last few years. The European Union (EU) has established strong restrictions on the commercial planting of genetically engineered plants, and European food processors and retailers are reluctant to import harvested agricultural products derived from biotechnology — thus jeopardizing the marketability of U.S. commodity grain exports. And the EU is now negotiating for strong restrictions on agricultural biotech products in important international agreements governing food safety and environmental protection. Very strong restrictions were included in the Biosafety Protocol, finalized in January 2000.

Perhaps, more important, though, are the Codex Alimentarius Commission standards for food safety. Without convincing scientific evidence that genetically engineered crop plants pose a heightened environmental or human health risk, restrictions on agricultural biotech imports could be challenged under the General Agreement on Tariffs and Trade and the World Trade Organization. Thus the EU has resorted to justifying its decisions on the basis of a risk-management philosophy known as the "precautionary principle," and the EU advocates inclusion of the precautionary principle in international environmental and food safety agreements (such as the Codex Alimentarius), as well as within GATT itself. Its inclusion would give nations greater latitude in restricting imports of U.S. agricultural products.

No single definition exists for the precautionary principle, but its general meaning is that when an activity raises threats of harm to human health or the environment, regulatory measures should be taken to prevent or restrict the activity even if the risks have not been demonstrated scientifically.

Although the EU asserts that the precautionary principle is an unbiased risk-management philosophy, critics have noted that its lack of definition and evidentiary standards makes it all too easy to abuse for the purpose of masking trade protectionism, and that its very approach to risk management is inherently flawed and may, in fact, increase net risk.. Nevertheless, the precautionary principle has already been included in several international environmental treaties, including the Convention on Biological Diversity and the Biosafety Protocol.

Even as underdeveloped nations clamor for biotechnology applications, and as countries like China continue to experiment with and use agricultural biotechnology, opponents of agricultural biotechnology in the West, particularly Europe, attack it as an unnatural process that will destroy the world, not better it. They argue that biotechnology should be heavily regulated, if not banned. Already, however, genetically engineered plants are subject to strict regulatory oversight that is equal to or greater than that advocated by the vast majority of scientific specialists. Additional regulation will slow down research and development of genetically engineered crops, keep beneficial products off the market, and raise the cost of products that do make it to consumers. Furthermore, the inclusion of similar restrictions — or inclusion of the precautionary principle — in international agreements will greatly impact the international trade of agricultural goods and delay their introduction into the marketplace. Each of these problems could prevent this technology's benefits from being introduced to industrialized nations and, more important, the developing world.

FUTURE OF AGRICULTURAL BIOTECHNOLOGY

There are divergent views about the rate of uptake of new agricultural biotechnologies and their impact. Kalter and Tauer and the US Office of Technology Assessment anticipate comparatively rapid take-up, while others (Buckwell and Moxey and Farrington do not foresee much impact until the next century, that is until ten to fifteen years have elapsed. The caution of the latter commentators is probably justified, for there are various hurdles to be jumped before biotechnologies can gain acceptance. One hurdle to be overcome is the economic one.

Self-evidently there must be some economic advantage to farmers or agro-industry from adopting the technology. Frequently what captures the scientific imagination turns out to be economically unviable in use, although it is only with use that efficiency can be improved and costs brought down. A more demanding test still for biotechnologies will be to obtain legislative approval and public acceptance. These are intimately connected.

In the case of machinery developments, provided safety standards are observed there are no legal impediments to developing new and improved machines, since there are no obvious problems of hazard to the public and therefore public acceptance. Chemical insecticide, fungicides and herbicides do, however, pose much greater problems because of concerns with toxicity, and elabourate testing and licensing procedures have evolved after a relatively haphazard set of procedures in the 1950s and 1960s, when widespread use was made of DDT and organo-phosphorous compounds without adequate recognition of their toxicity.

In significant part, because of the problems which have been experienced with agro-chemicals, the licensing of biotechnologies will inevitably be based on testing at least as stringent as that which now exists for chemicals. Indeed the licensing procedures are likely to be more stringent because of public and scientific concerns. Public acceptance of biotechnology in the food chain will be made more difficult by confusion about differences between biotechnologies. In the early 1980s there was a crescendo of concern about the use of steroids to promote faster liveweight gain in calves and cattle.

The public outcry eventually led to the banning of such hormones for meat animals; but the legacy is a profound distrust of and hostility to new products such as synthetic Bovine Somatrophin (BST), which has the capacity to increase the milk yield of those dairy cows which are 'relatively' deficient in what is a naturally occurring hormone. Although initial scientific results are favourable, and some minor doubts remain, the major influence leading the European Commission to impose a moratorium on the use of BST until the end of 1990 at least is concern about public perception; the Commission has stated 'It would be a serious setback to producers and to the Community's milk policy were present [positive] trends in consumption to be reversed as a result of adverse consumer reaction. With newly bioengineered plants and animals scientific and public concerns are emerging which are of a different order from those associated with previous biotechnology.

One relates to the possibility that crop failures may become more frequent if biotechnology leads to a reduction in genetic diversity in crops being grown; such a tendency has already been observed in relation to the Green Revolution technology for wheat. There are also concerns about upsetting the balance of nature in unforseen ways, akin to the unanticipated consequence of introducing rabbits into Australia and then trying to control these by introducing myxomatosis.

Thus questions arise such as what happens if herbicide resistance introduced into a commercial crop plant transfers itself by cross pollination into a closely related weed species, or indeed if the herbicide-resistant crop should colonize wild habitats. There are, of course, more lurid and absurd notions bandied about in the popular press which are similar in nature to the attacks made on Darwinism. The questions, both absurd and real, raised in relation to agricultural biotechnology will undoubtedly slow the rate at which it is adopted. Nevertheless instances of adoption are increasing: BST in the USA, a bioengineered baker's yeast in the UK, bioengineered sheep producing insulin, etc. As these become more widespread and increasingly affect lower-valued, bulk agricultural products, so the impacts of biotechnology on structural change will increase. The balance of economic power will switch increasingly to industry, to high-technology large farms and against smaller farmers in disadvantaged regions and countries. That, unfortunately, is a seemingly inevitable consequence of what we consider to be economic progress.

2

Agricultural Biotechnology of Food Science

FOOD BIOTECHNOLOGY

We have used biotechnology to manufacture food products for more than 8,000 years. Bread, alcoholic beverages, vinegar, cheese and yogurt, and many other foods and their existence to enzymes found in various microorganisms. Today's biotechnology will continue to affect the food industry by providing new products, lowering costs and improving the microbial processes on which food producers have long relied.

Many of these impacts will improve the quality, nutritional value and safety of the crop plants and animal products that are the basis of the food industry. In addition, biotechnology offers many ways to improve the processing of those raw materials into final products: natural flavors and colors; new production aids, such as enzymes and emulsifiers; improved starter cultures; more waste treatment options; "greener" manufacturing processes; more options for assessing food safety during the process; and even biodegradable plastic wrap that kills bacteria.

Improving the Raw Materials

The first generation of transgenic crops primarily benefited farmers. Although there are consumer benefits in growing these crops, the benefits are largely invisible to consumers. For example, studies have shown that because insect-resistant corn (Bt corn) sustains relatively little insect damage, fungi and molds cannot infect those plants as easily as non-insect-resistant crops. Therefore, the level of toxins, such as aflatoxin, produced by these pathogens, some of which are fatal to livestock, is much lower in Bt corn than non-Bt corn. The benefits of the next wave of biotechnology crops will be more obvious to consumers. Some of those benefits will involve improvements in food quality and safety, while others will provide consumers with foods designed specifically to be healthier and more nutritious.

HEALTH AND NUTRITIONAL BENEFITS

A variety of healthier cooking oils derived from biotechnology are already on the market. Using biotechnology, plant scientists have decreased the total amount of saturated fatty acids in certain vegetable oils. They have also increased the conversion of linoleic acid to the fatty acid found mainly in fish that is associated with lowering cholesterol levels. Another nutritional concern related to edible oils is the negative health effects produced when vegetable oils are hydrogenated to increase their heat stability for cooking or to solidify oils used in making margarine. The hydrogenation process results in the formation of trans-fatty acids.

Biotechnology companies have given soybean oil these same properties, not through hydrogenation, but by using biotechnology to increase the amount of the naturally occurring fatty acid, stearic acid. Animal scientists are also using biotechnology to create healthier meat products, such as beef with lower fat content and pigs with a higher meat-to-fat ratio. Other health and nutritional benefits of crops improved through biotechnology include increased nutritional value of crops, especially those that are food staples in developing countries.

Scientists at Nehru University in New Delhi used a gene found in the South American plant amaranth to increase the protein content of potatoes by 30 per cent. These transgenic potatoes also contain large amounts of essential amino acids not found in unmodified potatoes. Other examples include golden rice and canola oil, both of which are high in vitamin A. The golden rice developers further improved rice with two other genes that increase the amount and digestibility of iron.

Biotechnology also promises to improve the health benefits of *functional foods*. Functional foods are foods containing significant levels of biologically active components that impart health benefits beyond our basic needs for sufficient calories, essential amino acids, vitamins and minerals. Familiar examples of functional foods include compounds in garlic and onions that lower cholesterol and improve the immune response; antioxidants found in green tea; and the glucosinolates in broccoli and cabbage that stimulate anticancer enzymes.

We are using biotechnology to increase the production of these compounds in functional foods. For example, researchers at Purdue University and the U.S. Department of Agriculture, created a tomato variety that contains three times as much of the antioxidant lycopene as the unmodified variety. Lycopene consumption is associated with a lower risk of prostate and breast cancer and decreased blood levels of "bad cholesterol." Other USDA researchers are using biotechnology to increase the amount of ellagic acid, a cancer protective agent, in strawberries.

PRODUCT QUALITY

We are also using biotechnology to change the characteristics of the raw material inputs so that they are more attractive to consumers and more amenable to processing. Biotechnology researchers are increasing the shelf life of fresh fruits and vegetables; improving the crispness of carrots, peppers and celery; creating seedless varieties of grapes and melons; extending the seasonal geographic availability of tomatoes, strawberries and raspberries; improving the flavor of tomatoes, lettuce, peppers, peas and potatoes; and creating caffeine-free coffee and tea.

Japanese scientists have now identified the enzyme that produces the chemical that makes us cry when we slice an onion. Knowing the identity of the enzyme is the first step in finding a way to block the gene to create "tearless" onions.

Much of the work on improving how well crops endure food processing involves changing the ratio of water to starch. Potatoes with higher starch content are healthier because they absorb less oil when they are fried, for example. Another important benefit is that starchier potatoes require less energy to process and therefore cost less to handle.

Many tomato processors now use tomatoes derived from a biotechnology technique, somaclonal variant selection. The new tomatoes, used in soup, ketchup and tomato paste, contain 30 per cent less water and are processed with greater efficiency. A 1?2 per cent increase in the solid content is worth $35 million to the U.S. processed-tomato industry.

Another food processing sector that will benefit economically from better quality raw materials is the dairy products industry. Scientists in New Zealand have now used biotechnology to increase the amount of the protein casein, which is essential to cheese making, in milk by 13 per cent.

Biotechnology also allows the economically viable production of valuable, naturally occurring compounds that cannot be manufactured by other means. For example, commercial-scale production of the natural and highly marketable sweetener known as fructans has long eluded food-processing engineers. Fructans, which are short chains of the sugar molecule fructose, taste like sugar but have no calories. Scientists found a gene that converts 90 per cent of the sugar found in beets to fructans. Because 40 per cent of the transgenic beet dry weight is fructans, this crop can serve as a manufacturing facility for fructans.

SAFETY OF THE RAW MATERIALS

The most significant food-safety issue food producers face is microbial contamination, which can occur at any point from farm to table. Any biotechnology product that decreases microbes found on animal products and crop plants will significantly improve the safety of raw materials entering the food supply. Improved food safety through decreased microbial contamination begins on the farm. Transgenic disease-resistant and insect-resistant crops have less microbial contamination. Biotechnology is improving the safety of raw materials by helping food scientists discover the exact identity of the allergenic protein in foods such as peanuts, soybeans and milk, so they can then remove them.

Although 95 per cent of food allergies can be traced to a group of eight foods, in most cases we do not know which of the thousands of proteins in a food triggered the reaction. With biotechnology techniques, we are making great progress in identifying these allergens. More importantly, scientists have succeeded in using biotechnology to block or remove allergenicity genes in peanuts, soybeans and shrimp.Finally, biotechnology is helping us improve the safety of raw agricultural products by decreasing the amount of natural plant toxins found in foods such as potato and cassava.

Food Processing

Microorganisms have been essential to the food-processing industry for decades. They play a role in the production of the fermented foods. They also serve as a rich source of food additives, enzymes and other substances used in food processing.

IMPROVING FOOD FERMENTORS

Because of the importance of fermented foods to so many cultures, scientists are conducting a lot of work to improve the microorganisms that carry out food fermentations. The bacterium responsible for many of our fermented dairy products, such as cheese and yogurt, is susceptible to infection by a virus that causes substantial economic losses to the food industry. Through recombinant technology, researchers have made some strains of this bacterium and other important fermentors resistant to viral infection.

We have known for years that some bacteria used in food fermentation produce compounds that kill other, contaminating bacteria that cause food poisoning and food spoilage. Using biotechnology we are equipping many of our microbial fermentors with this self-defense mechanism to decrease microbial contamination of fermented foods.

FOOD ADDITIVES AND PROCESSING AIDS

Microorganisms have been essential to the food industry not only for their importance as fermentors, but also because they are the source of many of the additives and processing aids used in food processing. Biotechnology advances will enhance their value to the food industry even further.

Food additives are substances used to increase nutritional value, retard spoilage, change consistency and enhance flavor. The compounds food processors use as food additives are substances nature has provided and are usually of plant or microbial origin, such as xanthan gum and guar gum, which are produced by microbes. Many of the amino acid supplements, flavors, flavor enhancers and vitamins added to breakfast cereals are produced by microbial fermentation. Through biotechnology, food processors will be able to produce many compounds that could serve as food additives but that now are in scant supply or that are found in microorganisms or plants difficult to maintain in fermentation systems.

Food processors use plant starch as a thickener and fat substitute in low-fat products. Currently, the starch is extracted from plants and modified using chemicals or energy-consuming mechanical processes. Scientists are using biotechnology to change the starch in crop plants so that it no longer requires special handling before it can be used. Enzymes produced by microbial fermentation play essential roles as processing aids in the food industry. The first commercial food product produced by biotechnology was an enzyme used in cheese making. Prior to biotech techniques, this enzyme had to be extracted from the stomach of calves, lambs and baby goats, but it is now produced by microorganisms that were given the gene for this enzyme.

The production of high-fructose corn syrup from cornstarch requires three enzymes, and those same enzymes are important in making baked goods and beer. Other enzymes are essential to the production of fruit juices, candies with soft centers, and cheeses. The food industry uses more than 55 different enzyme products in food processing. This number will increase as we discover how to capitalize on the extraordinary diversity of the microbial world and obtain new enzymes that will prove important in food processing.

FOOD SAFETY TESTING

In addition to the many ways biotechnology is helping us enhance the safety of the food supply, biotechnology is providing us with many tools to detect microorganisms and the toxins they produce. Monoclonal antibody tests, biosensors, polymerase chain reaction (PCR) methods and DNA probes are being developed that will be used to determine the presence of harmful bacteria that cause food poisoning and food spoilage, such as *Listeria* and *Clostridium botulinum*.

We can now distinguish *E. coli* 0157:H7, the strain of *E. coli* responsible for several deaths in recent years, from the many other harmless E. coli strains. These tests are portable, quicker and more sensitive to low levels of microbial contamination than previous tests because of the increased specificity of molecular technique. For example, the new diagnostic tests for *Salmonella* yield results in 36 hours, compared with the three or four days the older detection methods required.

Biotechnology-based diagnostics have also been developed that allow us to detect toxins, such as aflatoxin, produced by fungi and molds that grow on crops, and to determine whether food products have inadvertently been contaminated with peanuts, a potent allergen.

ACCESSORY FOOD ITEMS

Although staple foods are generally used as vehicles in food fortification programmes, at times when none can be identified which has all the required characteristics, it has been necessary to find other options. In outlining fortification policy in the Philippines, Parce states that condiments may be fortified but only with nutrients that are deficient in the diet and provided that such food is an appropriate vehicle for the micronutrient and is widely consumed by the general population or is intended for intervention programmes to address deficiency in a specific target population.

SALT

Salt iodisation began in 1922 in Switzerland and has been implemented in many countries as the major mechanism for eliminating iodine deficiency. Today, IDD remains a problem in many countries. WHO and UNICEF have established the goal of 'Universal Salt Iodisation' to be achieved by the end of 1995. Salt has been favoured as a carrier for iodine due to its wide spread coverage, effectiveness, simple technology involved and low cost. Based on the suitability of salt as a widely used and low cost vehicle, fortification of salt with other nutrients has also been attempted.

According to the Codex standard for food grade salt, use can be made of potassium or sodium iodides and iodates. The iodates have been found to be more stable than the iodides under a wide range of conditions. Stability studies of iodised salt using potassium iodate as the fortificant demonstrated that there was no significant loss of iodine on storage in polyethylene bags for up to two years and that boiling of the salt solutions led to negligible iodine loss.

There have been four major technologies used in the addition of iodine to salt. These are dry mixing, drip feed addition, spray mixing and submersion: the salient features of these procedures have been described by Mannar. A thorough review of salt iodisation technology and quality control has been included in the recent text by Mannar and Dunn.

According to Bauernfeind, an acceptable iron fortificant for salt is one which does not discolour the salt nor impart a flavour or odour and remains stable and bio-available on storage. Ferrous fumarate, ferric orthophosphate and ferrous pyrophosphate have all been recommended by INACG for use with salt. Cook and Reuser report that the use of sodium acid pyrophosphate, ferric orthophosphate and ferric pyrophosphate as unacceptable due to poor bioavailability.

To circumvent this problem, iron absorption enhancers, such as ascorbic acid and sodium acid sulphate, have been used. Another possibility is to use less stable iron compounds in conjunction with a stabiliser. Such combinations include ferrous sulphate/hexametasulphate and ferrous sulphate/sodium acid pyrophosphate. Sodium ferric EDTA provides adequate amounts of iron, but has been found to discolour refined salt.

In the fortification of salt with iron, it must first be ground to a coarse powder to facilitate uniform mixing and distribution of the iron fortificant.

Rao reported that a ribbon blender or a screw type conveyor mixer with a suitable feeder could be used. He also reported that packaging in unlaminated high density polythene bags was found to be satisfactory.

The addition of both iodine and iron fortificants to salt can lead to the decomposition of the iodine compound and its subsequent liberation. Nestel reported on the use of a fortificant mixture containing 40 ppm of potassium iodate, and 1000 ppm of iron as ferrous sulphate and 10,000 ppm of a permitted stabiliser which rendered good bioavailability of both iodine and iron after prolonged storage. Cost was identified as a constraint in the use of this fortificant system as it added 50 per cent to the retail price of salt.

Fortification of salt with vitamin A has been attempted under laboratory conditions. The fortificant used was dry vitamin A palmitate type 250 SD protected by a lipid. The fortificant was found to be unstable at moisture contents above 2 per cent, since salt is hygroscopic, packaging material with an adequate moisture barrier must be used. Impurities in the salt were also found to destabilise the vitamin A. The particle size and shape must be such that uniform mixing could be achieved and segregation does not occur on storage.

MONOSODIUM GLUTAMATE

MSG is a condiment which is widely used in many Asian countries. It has GRAS status in Codex Alimentarius as a flavour enhancer. This additive has been thoroughly scrutinised due to the allergic response which it has been alleged to induce in susceptible individuals. Recent studies have indicated that it is safe for the vast majority of people. It is worthy of note that the Joint Expert Committee on Food Additives concluded that MSG should not be used in infant foods even though there was no cause for concern about health risks.

Field studies on MSG fortification with vitamin A have been conducted in the Philippines and Indonesia. In the Philippines dry vitamin A type 250 SD was used. This is vitamin A palmitate stabilised in an acacia-lactose matrix. The MSG was first ground to 100 mesh to facilitate mixing with the fortificant. Problems were encountered with segregation of fortificant and carrier, loss of vitamin A activity and colour deterioration of the fortified product. Vitamin A acetate 325 L, a granular fortificant stabilised in a gelatin - sucrose matrix, was used in place of the finely powdered vitamin A palmitate 250 SD. Under conditions of high humidity, there was hardening of the gelatin coat with associated loss of vitamin potency.

In the Indonesian study, dry vitamin A palmitate type 250 CWS dispersed in an edible carbohydrate, stabilised with antioxidants and coated with a white protective layer, was used. This product was hot and cold water miscible. This fortificant was demonstrated to retain over half of its potency when stored at 25 °C for 18 months. Under moist dark conditions, half of the activity is retained after 7 months. When subjected to light the fortificant is destabilised more rapidly, the uncoated form of the vitamin is less stable under these conditions. The fortificant mixture was added to MSG at the rate of 0.171 wt. per cent. The vitamin A was aggregated into clusters so as to minimise the problem of segregation during mixing and storage. Segregation of the product was still, however, identified as a problem. Colour deterioration of the fortified product was another constraint to the continued application of this technology.

Iron fortification of MSG has also been attempted using micronised ferric orthophosphate and zinc stearate coated ferrous sulphate. The coated ferrous sulphate had reduced bioavailability relative to the uncoated form, but the fortified product was judged to have acceptable colour, taste, bioavailability and particle size properties. Preliminary investigations indicated that the inclusion of dry stabilised vitamin A (type 250 SD) into the iron fortified mixture might be technically feasible.

SUGAR

Sugar has been found to be a suitable vehicle for nutrients in fortification programmes in Latin America and the Caribbean. In the vitamin A fortification of sugar, vitamin A 250-CWS was proven to be the most effective fortificant. The premix is produced by mixing sugar and fortificant in a revolving drum mixer. Physical separation of the fortificant beadlets from the sugar was prevented by the addition of an edible bonding agent. Stabilisers are added by first dissolving in warmed oil under anaerobic conditions, achieved by bubbling nitrogen through the oil, followed by intermittent addition to the fortified sugar with continuous mixing. Peanut, cotton seed, soya bean and coconut oils have all been used in this procedure. It has been suggested that shark liver oil be used in place of the vegetable oils.

The natural vitamin A content of this source reduced the need for vitamin A fortificant by 10-12 per cent. The successful use of shark oil would be expected to require rigorous deodorising and refining treatments and stabilisation of polyunsaturated fatty acids. Satisfactory storage stability of the premix was obtained when packaged in double polyethylene bags with an outer paper layer, and stored in a cool, dry place.

The premix was added to the sugar such as to achieve vitamin A levels of 50 IU per g in the final product. Addition of fortificant can take place in the centrifuge at the end of the washing cycle in the sugar refining process, or along the transporting belts prior to packaging. Retention of vitamin A after 6 months at 25 °C was 92 per cent. Higher losses were recorded with storage at 45 °C, with 76 per cent retention of vitamin activity after 3 months.

Fortification of sugar with iron has also been attempted. Bauernfeind reported promising results using sodium ferric EDTA as the fortificant. Segregation of the fortificant and the carrier was not a problem as the iron compound became stuck to the sugar crystals at moisture contents exceeding 1 per cent. A major problem which exists with iron fortified sugar is that on addition to coffee or tea, there is marked discolouration. This phenomenon is reduced with the use of NaFeEDTA, but it is still evident.

SAUCES

Iron fortification of strongly flavoured or dark coloured food products is often simplified due to the fact mat less care is required in consideration of the avoidance of off-flavours and off-colours. Largely for this reason iron fortification of curry powder demonstrated no technical difficulties. Nestel reported that NaFeEDTA was the fortificant used in the enrichment of curry powder in South Africa.

In the fortification of fish sauces the major problem was the formation of a precipitate on addition of iron. When sodium iron EDTA was the fortificant, however, this phenomenon was greatly reduced. The inclusion of iron in an EDTA complex greatly restricted its availability for interaction with macro-molecules and other compounds in foods which could eventually lead to precipitation or other deteriorative reaction.

FOOD PRODUCTION TECHNOLOGY

MECHANIZATION, INCOME, AND EROSION

Mechanization, which often came with the green revolution technologies, was said to be bad because it displaced labor. Again, it often happened that the increase in crops and output per year because of varieties that needed shorter growing seasons and higher yields per crop had labor requirements that more than offset the labor saved by mechanization.

In most areas, the extra crop, thereby demanding greater labor, would have been impossible without mechanization and the green revolution technologies. In the 1970s, saving labor by mechanization was always assumed to be a bad thing, though it might also mean more leisure for the farmer. Increasingly, mechanization means that farmers may earn cash income off the farm, affording school fees for children or consumption of products in the market economy.

It was conceded that the demand for agricultural labor was up, but that mechanization, by doing the more menial tasks, displaced women and children from the farm labor force. Many who oppose child labor in factories somehow deem agricultural stoop labor by children in the hot sun to be virtuous.

As incomes go up and child labor is displaced, there is a very good chance that they will now be in school. And where incomes go up, other technologies are also acquired that improve the quality of life and save on child labor. The high-yielding varieties of rice have been basically for production in irrigated fields. The upland varieties have barely changed in yields over the last three decades. This has led to the following argument:

The Green revolution has given rise to greater regional inequality of income. Careful empirical studies of rice cultivation in Asia have found that laborers from the upland regions migrate to work in the paddies, smoothing out the potential regional differences.

There are also arguments seeking a reduction in genetic variety, soil erosion, or groundwater contamination, which are of serious concern but cannot negate the enormous gain in human nutrition or the human catastrophe that would result from any attempt to reverse the green revolution. The diminution of genetic diversity is not a phenomenon of the green revolution per se but of agriculture. The individual farmer will select for yield or other output concerns, not for global genetic diversity. Modern no-tillage (or reduced or minimum tillage) agriculture using pesticides for weed and pest control conserves both soil and water better than its organic competitors.

MODERN FOOD SUPPLY AND SAFETY

Regularized and improved food supply has been a consistent factor throughout history in improving human health and longevity. In more recent times, improved food safety has added another dimension to human health. The evidence on the safety of our modern food supply is quite contrary to the strongly held conventional wisdom. Humans have had to deal with food contamination throughout history. Human foodstuffs have been the carriers of botulism, ergot, and the aflatoxins from the fungus *Aspergillus flavus,* which has caused mass illness, blindness, and largescale death. Even uncontaminated natural foods contain substances that would be considered a threat to human life if they were used as a food additive.

In the United States, during the early twentieth century, contaminated food, milk, and water caused many foodborne infections, including typhoid fever, tuberculosis, botulism, and scarlet fever. Once the sources and characteristics of foodborne diseases were identified, there were technologies that could help to control them before vaccines or antibiotics: "handwashing, sanitation, refrigeration, pasteurization, and pesticide application." Healthier animal care, feeding, and processing also improved food supply safety.

PLANT-PRODUCED CHEMICALS

The buzz word "chemicals," used by some to label and condemn food additives, could appropriately be used to describe what the plants themselves manufacture. Plants of all kinds, and most other life forms, are chemical factories, manufacturing a variety of chemicals, some of which have very active properties. Humans have extracted some of these chemicals for medicinal uses, some as poisons. Given the truism that dose makes the poison, some chemicals from plants that were once used as poisons are now used for medical purposes.

In other words, chemicals from plants are not inherently good or bad; goodness or badness is determined by a variety of factors and the uses to which they are put. One of the reasons for maintaining as much biodiversity as possible is that the earth's plants produce an array of active chemicals, the vast majority of which have not been screened for possible human use.

According to Katharine Milton, plants are unable to run from hungry predators and so have developed a variety of defenses to avoid the loss of edible components. "These protections include a vast array of chemicals known as secondary compounds (such as tannins, alkaloids and terpenoids). At best these chemicals taste awful; at worst, they are lethal".

Leopold and Ardrey add that these plant secondary substances are "chemicals that do not participate in the basic metabolism of the plant. Among these are many chemicals that serve to repel or discourage the use of the plant by insects, microorganisms, nematodes, grazing animals, and man." These plant toxins apparently are often used to ward off fungi.

"While antifungal chemicals in nightshade were distasteful ... there are probably protective compounds in other wild fruits that prevent rotting without spoiling taste, compounds that should be of particular interest to plant breeders". Many plants produce "phytoestrogens," the so-called estrogen disrupters, which might have the effect of reducing the fertility of their predators, the animals that graze or otherwise consume them. "Essentially, they give their predators an oral contraceptive".

Eating fresh, uncooked produce may have many taste and health benefits, but it can also be dangerous unless proper precautions are taken. As epidemiologist Michael T. Osterholmm stated: "The hearthealthy diet may be kind to our cardiovascular system but it is hell on the digestive tract".

Rachel Carson, whose book *Silent Spring* is seen as the inspiration of the modern environmental movement, recognized the role of natural toxins—systemic pesticides—in plants. She maintained that the perception of nature as an "enchanted forest" was a "fairy tale". Even some of the weeds associated with cultivars and often eaten with them are endowed with toxins and other substances deleterious to human health. Garn notes, "Weeds growing along with cultivars often included plants with interesting alkaloids, many of them in the tomato family that includes *Datura, Belladonna,* and the like".

TOXINS IN PLANT EVOLUTION

Through time, plants evolved for their own survival, not to serve human needs. "Plants in nature synthesize toxic chemicals in large amounts, apparently as a primary defense against the hordes of bacterial, fungal, and insects and other animal predators *plants in the human diet are no exception."*

The evolutionary process involved in domestication changed the survival strategy for these plants, as they now depended upon the continued cultivation to survive. The toxicity of many plants was reduced compared to their wild progenitors, but in most cases was not completely bred out of plants. In fact, currently there is considerable effort to breed plants with greater disease resistance so as to reduce the need for pesticides, in some instances by increasing the production of these toxins; in other instances, not.

GREEN REVOLUTION PROS AND CONS

It is hard to imagine anyone opposing the green revolution. The increase in yields from the green revolution in rice alone has produced enough to feed one billion people. Growth in yields accounts for 92 per cent of the increase in world cereal production since 1960. The world average for grain yields per hectare rose from 1.1 tons in 1950 to 2.9 tons in 1992. Without the increase in yields, it would have been necessary to bring another 3.6 billion hectares of land under cultivation, which would have almost doubled world cropland from "34 per cent of the earth's surface to 61 per cent".

The green revolution was in many ways a grain revolution— wheat, rice, and maize—but it also was a research revolution that facilitated an extraordinary and sustained general expansion in food production. Bender and Smith note, "Between 1961 and 1994, the number of daily food calories per capita rose from about 1,900 to 2,600 in developing countries, while their populations nearly doubled from 2.2 billion to more than 4.3 billion." Globally in the same period, "average daily per capita food supplies increased more than 20 per cent." The increase in available calories per capita for the developing countries rose 50 per cent from 1948–1952 to 1994–1996. A century-long trend of falling real food prices continued during the period 1950 to 1992 as international food commodity prices dropped 78 per cent in constant 1990 prices.

D. Gale Johnson sums up these achievements in agriculture: "People today have more adequate nutrition than ever before and acquire that nutrition at the lowest cost in all human history." In the 1990s, while population continued to increase, the absolute number of malnourished people in the world declined by forty million, which is a continuation of a trend of the last decades. However, this was considered a "failure" by the UN World Food Programme because it fell short of the reductions that were thought to be possible. To the extent that it is a failure, it is a failure of institutions, not of the technology that is producing the food that could make the programme a success.

So, there are arguments critical of the green revolution. The first argument is that it was a failure. Data such as that just cited above have demolished that argument, although a few still assert it, unaware of the change in the antitechnology party line. Next, we are told that the new varieties may have increased yields, but they led to a worsening of income distribution and the condition of the very poor. As early as the 1970s, studies emerged showing that the poor consumers and farmers benefited proportionately more than other groups as part of the overall improvement in human wellbeing.

Some argued that tenant farmers in the Punjab in the north of India were displaced from their land due to high-yielding varieties of wheat. It turns out that they returned to the land as farm laborers and earned more than they did as tenant farmers. Similarly, in southern India, between 1973 and 1994, the average real income of small farmers increased by 90 per cent, and that of the landless— among the poorest in the farm community—by 125 per cent. Further, caloric intakes for small farmers and the landless rose 58 to 81 per cent, and protein intakes rose 103 to 115 per cent. In addition, greater farm income meant more income and employment in the local villages that supplied the farm with goods and services now affordable to those who worked the land.

The green revolution in Indian agriculture is credited with being a major force in poverty reduction in India. From independence to 1970, the per centage of the population of India living below the absolute poverty line remained relatively steady at 55 per cent. In over two decades since the introduction of the green revolution technologies, the per centage of the population of India living in absolute poverty has fallen to 35 per cent. Throughout most of East and Southeast Asia, the early phases of development involved an emphasis on agriculture and food production.

The green revolution technologies in agriculture have played an important role in the dramatic reductions in poverty rates throughout the region, such as in Indonesia from 58 per cent in 1970 to 8 per cent in 1993, and in China from 270 million people in poverty in 1978 to 65 million in 1999. Yao estimates the reduction in rural poverty in China to be from between 596 to 790 million people (75.5 per cent–100 per cent) in 1978 to 57 to 114 million (6.7 per cent–13.2 per cent) in 1996 (447). By either measurement, the magnitude of the reduction in poverty in China is extraordinary and without precedent in human history.

In the time period 1969–1971 to 1990–1992, global population was increasing about 45 per cent while the absolute number of people suffering from chronic undernourishment in developing nations decreased from 917 million or 35 per cent of their population to 839 million or 21 per cent. From 1990–1992 to 1995–1997, the absolute number of undernourished people in the developing world continued to decline, falling to 790 million.

The prevalent high number should not be minimized, nor should we be complacent, but neither should the progress achieved be ignored. For we can finish the job of eliminating undernourishment by recognizing the forces that have brought us this far, thereby gaining some understanding of what still needs to be done. Unfortunately, many wish to deny the progress that has been made, offering antitechnology solutions, which would reverse these gains.

BIOTECHNOLOGY AND FOOD PRODUCTION

Scientific innovation and its derivative benefits have had profound implications to humanity within the last century. The exciting discipline of biotechnology has drawn the interests of traditional biologists, biochemists, microbiologists, medical and agricultural scientists into applying mathematical and engineering models to understanding biology. Furthermore, several scientists in the exact sciences of mathematics, physics, and chemistry have begun to use system approaches to unravel the mystery and complexity of biology. And from the side, diagnostic, biopharmaceutical, biochemical and agriculcultural industries are rapidly drawing from and applying the research results of biotechnology. Biotechnology is experiencing a revolution like none before in the life sciences and is affecting every facet of our lives, from crop improvement to commerce, and drugs to sustainable development. This technology has the potential to address problems not solved by conventional research.

At the same time, biotechnology may speed up research process and increase research precision. GM crops have been developed and rapidly disseminated since the early 1990s. At present, there is very little commercial utilization of results from modern biotechnology research in developing countries. As a result, the potential contributions of biotechnology to poverty alleviation, enhanced food security and nutrition in developing countries have received little attention, beyond blanket statements of support or opposition.

THE GENE REVOLUTION

Mendel's laws of genetics were rediscovered in 1900. Mendel had published his work on inheritance patterns in pea in 1865, but it took 35 years for others to grasp their significance. Since 1900, there has been steady progress in our understanding of the genetic makeup of all living organisms ranging from microbes to man. A major step in human control over genetic traits was taken in the 1920s when Muller and Stadler discovered that radiation can induce mutations in animals and plants.

In the 1930s and 1940s, several new methods of chromosome and gene manipulation were discovered, such as the use of colchicine to achieve a doubling in chromosome number, commercial exploitation of hybrid vigour in maize and other crops, use of chemicals such as nitrogen mustard and ethyl methane sulphonate to induce mutations and techniques like tissue culture and embryo rescue to get viable hybrids from distantly related species. The double helix structure of DNA, the chemical substance of heredity, was discovered in 1953 by James Watson and Francis Crick. This triggered explosive progress in every field of genetics. Just as it took 35 years for biologists to understand the significance of Mendel's work, it may take a couple of decades more to understand fully the benefits and risks associated with genetically improved foods. It would be prudent to apply scientific and precautionary principles in areas of human health and environmental safety. The 1990s have seen dramatic advances in our understanding of how biological organisms function at the molecular level, as well as in our ability to analyse, understand, and manipulate DNA molecules, the biological material from which the genes in all organisms are made. The entire process has been accelerated by the Human Genome Project, which has poured substantial resources into the development of new technologies to work with human genes.

The same technologies are directly applicable to all other organisms, including plants. Thus, the new scientific discipline of genomics has arisen, which has contributed to powerful new approaches in agriculture and medicine, and has helped to promote the biotechnology industry. The quality of food and food plants can be modified and optimized to meet the nutritional and health needs of atrisk and compromised populations prevalent in most of the developing countries. High rates of malnutrition, infectious disease as well as diet-related diseases such as diabetes and hypertension are prevalent in many developing countries.

These are as a result of compromised immune function, inadequate sources of nutritious and quality foods and limited access to healthy and suitable foods. Biotechnology and genetic modification techniques have been proposed and applied for the improvement of the quality of various food crops.

These have typically been geared towards increasing yields and pest resistance of cash crops. Furthermore, the application of biotechnology techniques for the development of functional food plants with higher levels of bioactive components or increased availability of nutrients would greatly benefit most populations in developing countries and improve the health and nutritional status overall. In certain areas, biotechnology and genetic modification techniques are being optimized for the production and development of healthy foods, and improvement in the levels and activity of biologically active components in food plants. Biotechnology techniques in developing countries however have mostly been targeted at increasing yields of cash crops. Food crops or the improvement of food quality and functional foods have garnered much less attention.

Techniques applied in genetic modification include mutation breeding, improved conventional breeding, transgenic modifications, DNA insertion, gene transfer and somatic hybridization. While there has been some hesitation with regard to the acceptability and adoption of biotechnology products in certain developing countries, achievements such as the development of high-vitamin A rice have greatly increased the acceptability of biotechnology for human food applications among hitherto skeptical consumers. Furthermore, it provides insights into the potential for application of biotechnology in developing improved quality and functional foods for human nutrition and health, rather than simply for use in agricultural technology for improved yields and pest resistance. The production of increased levels of beta-carotene in plants is especially important, as its precursor, lycopene has been shown to have physiological chemo-preventive effects with regard to various cancers. Furthermore, lycopene, commonly found in various carotenoid containing plants such as tomatoes and carrots, is an essential ingredient in maintaining eye health and vision. Modifications that have been targeted and developed by various biotechnology companies include improvement in the oil content and composition of oil seeds such as legumes. Improvement in soybean oil quality includes stabilization of the unsaturated fatty acids by increasing levels of the antioxidant, vitamin E. These successes indicate a relevant and important role for biotechnology in improving food quality and developing functional foods, particularly those targeted for needy populations in developing countries, such as children and pre-natal women.

GENETICALLY MODIFIED CROPS

Modern agricultural biotechnology is still in an early phase, and the focus is overwhelmingly on production in industrial country farms and for industrial country markets. In 1998, 85% of the land planted with genetically improved crops was in just five developed countries with the United States alone accounting for about 75% of the area. Argentina, China, Mexico, and South Africa cultivated the remaining 15%, and the countries other than China include a substantial number of large-scale, capital-intensive farms that produce primarily for industrial country markets.

Among the crops produced in these four developing countries are insectresistant cotton and maize, herbicide-resistant soybean, and tomatoes with a long shelf life. Globally, herbicideresistant soybean, insect-resistant maize, and genetically improved cotton account for 85 per cent of all plantings.

The unprecedented rapid adoption of transgenic crops during the initial five-year period when genetically modified crops were first adopted, reflects the significant multiple benefits realised by large and small farmers in industrial and developing countries that have grown transgenic crops commercially. Between 1996 and 2000, a total of fifteen countries - 10 industrial and 5 developing - contributed to more than a twenty five fold increase in the global area of transgenic crops from 1.7 million hectares in 1996 to 44.2 million hectares in 2000.

The accumulated area of transgenic crops planted in the five-year period 1996 - 2000 total 125 million hectares, equivalent to more than 300 million acres. Adoption rates for transgenic crops are unprecedented and are the highest for any new technology by agricultural industry standards. High adoption rates reflect growing satisfaction with products that offer significant benefits ranging from more convenient and flexible crop management, higher productivity and/or net returns per hectare, health benefits and a safer environment through decreased use of conventional pesticides, which collectively contribute not only to improved weed and insect pest control but also benefits of lower input and production costs; genetically modified crops offer significant economic advantages to farmers compared with corresponding conventional crops. The severity of weed and insect pests varies from year to year and hence this will have a direct impact on pest control costs and the consequent economic advantage.

Despite the on-going debate on GM crops, particularly in countries of the European Union, millions of large and small farmers in both industrial and developing countries continue to increase their plantings of GM crops because of the significant multiple benefits they offer. This high adoption rate is a strong vote of confidence in GM crops, reflecting grower's satisfaction. Many recent studies have confirmed that farmers planting herbicide-tolerant and insect-resistant Bt crops are more efficient in managing their weed and insect pests.

An estimated 3.5 million farmers grew transgenic crops to health and economic advantages. In coming years, the number of farmers planting GM crops is expected to grow substantially and the global area of GM crops is expected to continue to grow. Global population would exceed 6 billion by 2050, when approximately 90% of the global population will reside in Asia, Africa and Latin America.

Today, 815 million people in the developing countries suffer from malnutrition and 1.3 billion are afflicted by poverty. Transgenic crops, often referred to as GM crops, represent promising technologies that can make a vital contribution to global food, feed and fibre security. During the six-year period 1996 - 2001, the global area of transgenic crops increased by more than 30-fold, from 1.7 million hectares in 1996 to 52.6 million hectares in 2001.

This high rate of adoption reflects the growing acceptance of transgenic crops by farmers using GM technology in both industrial and developing countries. During the six-year period 1996 - 2001, the number of countries growing transgenic crops more than doubled, increasing from 6 in 1996 to 9 in 1998, to 12 countries in 1999 and 13 in 2000 and 2001. The experience of the first six years, 1996 to 2001, during which a cumulative total of over 175 million hectares of transgenic crops were planted globally in 16 countries, 10 industrialized and 6 developing, has vindicated the vision of the pioneers of crop biotechnology who have seen their early promises of transgenic crops fulfilled - GM crops have met the expectations of large and small farmers planting transgenic crops in both industrial and developing countries. The unprecedented rapid adoption of transgenic crops during the first six-year period, 1996 - 2001, reflects the significant multiple benefits realised by large and small farmers.

There is a growing body of evidence that clearly demonstrates the improved weed and insect pest control attainable with transgenic herbicide-tolerant and insect resistant Bt crops that also benefit from lower input and production costs. Despite the ongoing debate on GM crops, particularly in countries of the European Union, millions of large and small farmers in both industrial and developing countries continue to increase their plantings of GM crops because of the significant multiple benefits they offer.

More specifically, the use of transgenic crops results in:

- More sustainable and resource-efficient crop management practices that require less energy and fuel and conserves natural resources.
- More effective control of insect pests and weeds.
- A reduction in the overall amount of pesticides used in crop production, which impacts positively on biodiversity, protects predators and non-target organisms and contributes to a safer environment.
- Less dependency on conventional pesticides that can be a health hazard to producers and consumers; the potential health benefits associated with fewer pesticide poisonings from Bt cotton in China is an important finding, with significant implications for other developing
- Countries where small farmers may be at similar risk from heavy and over-use of conventional pesticides.
- Bt maize, which has reduced levels of the fumonisin mycotoxin provides safer and healthier food and feed products.
- Greater operational flexibility in timing of herbicide and insecticide applications.
- Conservation of soil moisture, structure, nutrients and control of soil erosion through no or low-tillage practices as well as improved quality of ground and surface water with less pesticide residues.

There is now considerable evidence that transgenic crops are delivering significant economic benefits. The global economic advantage to farmers deploying herbicide tolerant soybean, HT canola and Bt corn was estimated to be of the order of $ 700 million in 1999, equally shared between developing and industrial countries. In addition to these direct economic advantages that farmers derive from transgenic crops, there are also significant additional indirect benefits to others in society. For crops such as herbicide-tolerant soybean, these indirect benefits to consumers globally can be of the same order of magnitude as the direct economic advantages to farmers.

Thus, the global direct and indirect economic advantage of GM crop in 1999 was of the order of $ 1 billion or more. 2001 is the initial year of the second quinquennium during which GM crops is being comercialized. In 2001, coincidental with increased political, policy and institutional support for GM crops due to their acknowledged contribution to global food security, the global area of transgenic crops benefited from renewed growth which resulted in a 19% increase in global GM crop area in 2001 - almost twice the growth rate in 2000. The number of farmers that benefited from GM crops increased from 3.5 million farmers in 2000 to an estimated 5.5 million in 2001. More than three-quarters of the GM farmers that benefited from GM crops in 2001 were resource-poor farmers planting Bt cotton, mainly in eight provinces in China and also in the Makathini Flats in KwaZulu Natal province in South Africa.

The well documented experience of China with Bt cotton presents a remarkable case study where 5 million small resourcepoor farmers in 2001 are already benefiting from significant agronomic, environmental, health and economic advantages - this is a unique example of how biotechnology can impact on poverty alleviation, as advocated in the 2001 UNDP Human Development Report. The China experience with Bt cotton lends itself for introduction and replication to carefully selected developing countries in Asia, Latin America and Africa where resource-poor farmers can learn, share and benefit from the rich experience of China – the majority of the hectare of global cotton is grown in developing countries. Following a successful launch of Bt cotton in 2001, Indonesia is also expected to expand its Bt cotton in 2002, and India, the largest cotton growing country in the world, has approved commercial cultivation of Bt cotton in 2002. There is cautious optimism that global area and the number of farmers planting GM crops will continue to grow in 2002 in the six principal countries already growing GM crops - USA, Argentina, Canada, China, South Africa and Australia.

The other seven countries growing transgenic crops in 2001 are expected to report modest growth in GM crop area in 2002. The commercialization of herbicide-tolerant soybean in Brazil will be dependent on resolving the outstanding regulatory issues between the Ministries of Agriculture, Environment and Justice. The commercialization of GM crops in India and Brazil would represent a watershed for GM crops in developing countries in that the three most populous countries in Asia - China, India and Indonesia with 2.5 billion people, as well as the three major economies of Latin America - Argentina, Brazil and Mexico, plus South Africa would then all be commercializing and benefiting from transgenic crops. As new and novel products with input and output traits will become available for commercialization in the near future, it is critical that these products be deployed in an integrated strategy in which both conventional and biotechnology applications are applied to attain the challenging goal of global food security. Adoption of such a strategy will allow society to continue to benefit from the vital contributions that both conventional and modern plant breeding offer. Technology is only one of several elements that can contribute to an integrated global food security strategy where population control and improved food distribution systems are also essential elements. Biotechnology can play its appropriate and essential role in achieving food security in the developing world. The experience of China, Argentina and South Africa, that are already deriving significant benefits from GM crops, can be shared with other developing countries in the three continents of the South which face similar challenges.

ADAPTING BIOTECHNOLOGY FOR INCREASED CROP PRODUCTION

What is needed in developing countries is an understanding that goes beyond conventional, orthodox wisdom to work more strategically in developing and implementing effective, international, national and regional policies. Funding for agricultural research has declined 50% on a worldwide basis. Funding for research in general is at a dismal low in developing countries.

The intrusion of intellectual property rights into the arena of crop improvement, while beneficial to the economies of the developed world is making the lives of many researchers even more difficult. Private industry has dominated research in biotechnology. Consolidation of the industry has proceeded rapidly since 1996, with more than 25 major acquisitions and alliances worth US$15 billion. Little private-sector agricultural biotechnology research so far has focused on developing country food crops other than maize. Production of genetically modified crops is not a complex technology and is clearly within the capabilities of national research institutes in many developing countries. Genetic modification of crops using recombinant DNA technology is also within reach of the institutes of the Consultative Group on International Agricultural Research, including Centro Internacional de Mejoramiento de Maiz y Trigo in Mexico, International Rice Research Institute in The Philippines, and the International Institute for Tropical Agriculture in Nigeria.

Furthermore, these institutes have already assumed responsibility for biotechnological research and a number of crop improvement projects are under way. These institutes see biotechnology as a tool and not as an end in itself. Crop improvement through biotechnology need not be equated with transgenic plants. For example, marker-assisted breeding is a powerful biotechnology that can find widespread application with the crops of the poor. Detailed linkage maps of these crops will be tremendously useful. As these CGIAR institutes focus on their needs, they will want and need to reach out to public institutions in developed countries. Alliances such as the Cassava Biotechnology Network that bring together researchers from many countries are an effective way to create synergy towards a common goal.

The private sector leads in every aspect of the agricultural biotechnology revolution and activities in the public sector will have to marshal the strength of the private sector through public-private partnerships. Such partnerships must be based upon mutual trust and common goals. The private sector can work with the CGIAR institutes and with national research institutions of developing countries to transfer technologies, train scientists, provide hands-on experience in intellectual property management, and facilitate the nocost or low-cost licensing of inventions. Since the cost of these inventions is being charged to the consumers in developed countries, such approaches amount to a transfer of wealth by large corporations from the developed world to the developing world.

The application of biotechnology to the problems of the poor will not be straightforward and the models we have from developed countries will probably not be applicable. Agricultural research for the crops and problems of the poor has to proceed from the bottom up, not from the top down. Crops have to be created that fit not only in the agro-ecology of the poorest regions often characterized by marginal and heterogeneous environments, but the crops must also fit into the social and economic systems. Agricultural research has to start with studying farming practices, asking the farmers - men and women - what they want, allowing the farmers to make choices between often conflicting objectives such as higher yield versus yield stability, and examining the possibility of marketing the excess production. Researchers have to begin by soliciting the help of the farmers to describe farming practices and analyse these practices to pinpoint problem areas and opportunities. Together, the researchers and farmers have to generate a range of choices that the farmers could implement. The major objective of this approach is not the transfer of technology, but empowerment of the farmer to improve production.

AGRICULTURAL BIOTECHNOLOGY AND FOOD SAFETY

When a consumer buys food s/he expects it to be safe and have the same quality characteristics that s/he is accustomed to, and familiar with. Consumers have the right to know that this is the case and if there is anything different about that product. The Food and Agriculture Organization and the World Health Organization of the United Nations advocate the concept of 'substantial equivalence' as the most practical approach to address the safety evaluation of foods or food components derived through modern biotechnology. This approach states that 'if a new food or food component is found to be substantially equivalent to an existing food or food component, it can be treated in the same manner with respect to safety.' So what kind of tests are used to determine if the food is as safe or safer than existing foodstuffs? Researchers must prepare comprehensive data to support the safety and wholesomeness of new crop varieties developed through biotechnology.

This process requires years of laboratory and field testing before a product can be brought to the market. To provide assurance that foods derived through biotechnology are as safe as those produced by traditional breeding programmes, the safety assessment strategies involve several key steps. These steps include molecular characterisation of the genetic modification, agronomic characterisation, nutritional assessment, toxicological assessment and safety assessment.

For example, typical questions that must be addressed are:

- Does the genetically modified food have a traditional counterpart that has a history of safe use?

- Has the concentration of any naturally occurring toxins or allergens in the food changed?
- Have the levels of key nutrients changed?
- Do new substances in the genetically modified food have a history of safe use?
- Has the food's digestibility been affected?
- Has the food been produced using accepted, established procedures?

Even after these and other questions about the biotechnology derived foods are answered, there are still more steps in the approval process before the crop can be commercialized. In fact, genetically modified foods are the most studied food products ever produced.

- *Safety assessment*: Molecular characterisation – for new plant varieties produced through modern biotechnology, the source of the gene introduced into the plant is first identified. The transformation system used to insert the gene is defined, as well as the number of copies of inserted genes and the integrity and stability of the genetic insert.
- *Agronomic traits*: Usually the starting points for evaluating substantial equivalence. For example, in the case of potatoes, the traits commonly examined are yield, tuber size and distribution, dry matter content and disease resistance.
- *Nutritional assessment*: Involves key nutrients including fats, proteins, carbohydrates and essential vitamins and minerals.
- *Toxicology assessment*: Toxicants and anti-nutrients are compounds known to be naturally present in some crops that could have an impact on health if their levels increased. For example, solanine glycoalkaloids in potatoes or trypsin inhibitors in soybeans). The levels of anti-nutrients in crops produced through biotechnology are compared to conventionally produced varieties grown under comparable environmental and agronomic conditions.
- *Safety assessment*: When a crop produced through biotechnology is shown to be substantially equivalent to a conventional crop, the safety assessment focuses on the introduced trait and the expressed protein product.

The overall goal of these tests is to determine whether the plant is substantially equivalent to food derived from a conventional source that has a history of safe use. A substantial equivalence evaluation focuses on the product rather than the process used to develop the product. If the new product is substantially equivalent to the conventional food or feed, then the product derived through biotechnology is considered to be as safe as the conventional counterpart. If the food produced using biotechnology contains a new trait, which changes the levels of nutrients or antinutrients, such as a higher level of a vitamin or a lower level of an allergen, the assessment focuses on demonstrating the safety of the new trait.

ALLERGENS

One of the public's biggest concerns related to genetically modified foods is that an allergen could be accidentally introduced into a food product. The DNA inserted into a plant is safe and no case of allergy to DNA is known, the concern is whether the protein produced by the plant as a result of the inserted DNA could cause an unexpected allergic reaction. Scientists know a lot about which foods trigger allergic reactions in adults and children. Ninety per cent of all food allergies are associated with only eight foods or food groups – shellfish, eggs, fish, milk, peanuts, soybeans, tree nuts, and wheat. These, and many other food allergens, are well characterized. To date, genes found to express any known allergenic protein have been specifically excluded from the research and development process.

Scientists have agreed that following these simple steps provides compelling assurance that no allergic effects will result from foods derived from biotechnology:

- Avoid transferring genes from foods known to be allergenic.
- Check the structure of any new proteins produced in foods derived from biotechnology against the structures of known allergens to assure that no allergenic structures exist in the new protein.
- Measure the stability of the new protein in stomach and intestinal fluids. Most allergens are stable to these conditions. Proteins which are unstable to these conditions are not likely to be allergens.
- Determine how much of the new protein will be present in the food consumed by humans. Most allergens are present in large amounts.

Allergenicity screening continues to be a very important part of safety testing before a crop can enter into the food market.

ANTIBIOTIC RESISTANCE

Some biotechnology crops contain genes for a trait called antibiotic resistance. Scientists use this trait as a marker to identify cells into which the desired gene has been successfully introduced. Concerns have been raised that these marker genes could move from biotechnology crops to microorganisms that normally reside in a person's gut and lead to an increase in antibiotic resistance.

There have been numerous scientific reviews and experimental studies of this issue and they have come to the following conclusions:

- The likelihood of antibiotic resistance genes moving from biotechnology crops to other organisms is extremely remote.
- In the unlikely event that an antibiotic resistance gene is transferred to another organism, the impact of this transfer would be negligible, as the markers used in genetically modified crops have limited or no clinical or veterinary use.

However, in response to public concerns, scientists have been advised to avoid using antibiotic resistance genes in biotechnology modified plants. Alternative marker strategies are being developed.

GAUGING UNDERSTANDING

The development of any educational material should be based on a sound understanding of what the audience currently knows. In 1998-99, AFIC commissioned market research to identify consumer knowledge and attitudes towards biotechnology in Asian countries. AFIC's research was undertaken in the urban areas of Singapore, Malaysia, the Philippines, Indonesia and Thailand. In 2002, the survey was repeated with additional questions to provide an updated and more in-depth understanding of consumer knowledge and attitudes with regard to food biotechnology.

The survey was done in the urban areas of China, the Philippines and Indonesia. The survey found that the majority of consumers are aware of the presence of biotechnology-derived foods in their everyday diet, and are not averse to this situation. Biotechnology foods rated as the issue of least concern while those of the greatest concern were nutritional value, microbial contamination and pesticide residues. In 2003, AFIC conducted research using focus group discussions methods in the Philippines, China, and India to assess the consumer perception towards biotechnology foods as well as to test and develop appropriate educational messages on issues relevant to food biotechnology. These countries were selected partly because they had progressed regulatory decisions on the cultivation of biotechnology in the previous 12 months and thus discussions could reference actual, rather than hypothetical national policy status.

The main findings of the studies are as follows:

- The majority of consumers adopted an open-minded position towards biotechnology foods.
- What consumers are most concerned about are the use of pesticides and preservatives in food production.
- Many participants in the research surveys clearly had very limited knowledge about biotechnology. However increasing knowledge levels were associated with increasingly positive acceptance of biotechnology foods.
- Survey participants expressed clear desire for more factual information in understandable language and format
- The application of biotechnology to potentially produce foods with enhanced nutritional value or requiring less pesticide for cultivation, elicited very positive responses.
- Consumers were very unaware of prevailing concerns being debated within the inner circle of stakeholders and were not seeking information on safety and risk assessment.
- Unsolicited information on safety assessment did not appear to improve knowledge levels, but instead generated anxiety, no matter how the information was presented.

AGRICULTURAL BIOTECHNOLOGY AND FOOD SECURITY

Can molecular biology-based research contribute to the solution of the problems outlined earlier? Are the potential social and economic benefits likely to exceed potential risks or costs? If these questions are answered in the affirmative, issues related to the design of the technology and the needed policies and institutions must be tackled.

Although conventional applications of biotechnology, such as tissue culture and fermentation amongst others, is under way in several developing countries, little genetically improved seed material has been grown in the poorer developing countries to date so ex post assessment is virtually impossible. A great deal is known, however, about the social and economic risks and benefits associated with traditional Mendelian plant breeding as exemplified by the Green Revolution.

The analysis, therefore, begins with the identification of similarities and differences between the Green Revolution and modern biotechnology, and an attempt is made to draw sessions from the Green Revolution and to look at the difference between that technology package and modern biotechnology to try to assess the likely social and economic risks and benefits of modern agricultural biotechnology ex ante.

COMPARING THE GREEN REVOLUTION AND MODERN BIOTECHNOLOGY

Shift to Private Sector Research

There are three differences of particular importance for an assessment of social and economic risks and benefits. The research leading to the Green Revolution was undertaken by the public sector and the improved seed was usually freely available for seed multiplication and distribution. Although breeders' rights may permit an initial charge for the improved materials, the intellectual property rights did not extend beyond the initial release.

Having acquired the seed, farmers could reuse it without further payment, although reuse of hybrid seed would drastically reduce the yield advantage. This is in keeping with the principle of "farmers' rights" included in the 1983 International Undertaking on Plant Genetic Resources.

In contrast, the bulk of modern agricultural biotech-nology research is undertaken by private sector firms, which protect IPRs through patents that extend beyond the first release.

Farmers, therefore, cannot legally plant or sell for planting the crop produced with the patented seed without the permission of the patent holder. Patent holders, currently seeking ways to enforce their rights, are considering approaches such as legal agreements and technologies that will activate and deactivate specific genes. However, monitoring and enforcing contracts that prohibit large numbers of small farmers from using the crops they produce as seed would be expensive and difficult. The so-called terminator gene is the first patented technology aimed at biological IPR protection. It is not appropriate for small farmers in developing countries because existing infrastructure and production processes may not be able to keep fertile and infertile seeds apart.

Small farmers could face severe consequences if they planted infertile seeds by mistake. Commercialization of the terminator gene now seems unlikely in the short term. Research is under way on other biological approaches to IPR protection that would not impose such risk on small farmers. These include, for example, genetically engineered seeds that contain desired traits, such as pest resistance or drought tolerance, but in which these are activated only through chemical treatment. Otherwise, the seed would maintain its normal characteristics. Thus, if a farmer planted an improved seed, the offspring would not be sterile; rather they would revert to normal seeds, without the improved traits. The farmer would have the choice of planting the seed and doing no more, or activating the improved traits by applying the chemical. This approach complies with the principle of doing no harm.

It is important to note that even when patents permit a private company to enjoy monopoly or near-monopoly rights over a product it has developed, the firm is unlikely to capture 100 per cent of the economic benefits. A recent study of the distribution of the economic benefits generated by the use of herbicide-tolerant soybean seed in the United States in 1997 found that the company, Monsanto, received 22 per cent, while seed companies gained 9 per cent. Consumers of soybean and soybean products in the United States and other countries reaped a 21 per cent share, whereas farmers worldwide obtained 48 per cent. The share of U.S. farmers was actually 51 per cent of the benefits, but farmers elsewhere experienced net losses of 3 per cent.

Rise of Proprietary Research Processes and Technologies

A second, and related, difference between the Green and Gene revolutions involves the patenting of processes as well as products. The main process behind the Green Revolution was conventional plant breeding technology, which lies in the public domain, carried out by public institutions. Today, the processes used in modern agricultural biotechnology are increasingly subjected to IPR protection, along with the products that result. This means that public sector research institutions may not be able to gain access to basic but proprietary knowledge and processes needed in research, including research on the so-called orphan crops such as cassava and millet.

These are critical staples in the diets of many poor people, but they do not offer promising economic returns to private sector R&D efforts, so efforts to develop disease-resistant cassava or drought-tolerant millet, whether through genetic modification or conventional breeding, must come from the public sector. Some firms have agreed to transfer proprietary technologies, without charging royalties, to developing countries where there are few potential commercial prospects. Monsanto, for example, has entered into agreements with Kenyan and Mexican government agricultural research institutes to develop virus-resistant crops. Arrangements such as these are few and generally involve the philanthropic arms of the private firms.

Enlightened Adaptation Vs. Direct Transfer

A third difference involves the adaptation of industrial country agricultural research to developing country conditions. Although based on earlier research in industrial countries, the Green Revolution was focused on solving specific problems in developing countries. Current application of modern biotechnology is focused on industrial country agriculture. Industrial country research institutions had begun working on development of higher yielding crop varieties in the late 19th century. For example, in Japan, rice breeding under the auspices of the Ministry of Agriculture and public universities led to large yield gains in the early part of the 20th century, with a second wave of major gains after 1945. During the early decades of Soviet history, under the leadership of geneticist Nikolai Ivanovich Vavilov, the government carried out extensive crop improvement programmes and established one of the world's largest germplasm collections. In the United States, hybrid maize research began in the 1920s.

Much of the basic research was done by public institutions, such as land grant universities, state experiment stations, and the U.S. Department of Agriculture. Applications to particular farming conditions and the mass marketing of the new varieties were, in turn, handled by private seed firms such as Pioneer Hi-Bred and DeKalb. The research focused not only on developing higher yielding seeds to bolster food supplies for domestic consumption but also on animal feed and production for export. This research could not simply be transferred to poorer developing countries, where the need was for improved varieties of locally-consumed staples.

The research that led to the Green Revolution involved further adaptation to the agroecological conditions of tropical and semitropical areas. It also focused on rice, wheat, maize, root and tuber crops, and tropical fruits and vegetables. The public sector role was, if anything, even more prominent, with international agricultural research centres and national agricultural research systems, particularly in Asia and Latin America, playing a prominent role. Financial support came from donors of official development assistance and large private foundations, such as Ford, Rockefeller, and Kellogg.

In contrast, modern agricultural biotechnology is still in an early phase, and the focus is overwhelmingly on production on industrial country farms and for industrial country markets. In 1998, 85 per cent of the land planted to genetically improved crops was in just five developed countries with the United States alone accounting for about 75 per cent of the area. Argentina, China, Mexico, and South Africa cultivated the remaining 15 per cent, and the countries other than China include a substantial number of large-scale, capital-intensive farms that produce primarily for industrial country markets. Among the crops produced in these four developing countries are insect-resistant cotton and maize, herbicide-resistant soybean, and tomatoes with a long shelf life. Globally, herbicide- resistant soybean, insect-resistant maize, and genetically improved cotton account for 85 per cent of all plantings.

Both the area planted to genetically improved crops and the value of the harvests grew dramatically between 1995 and 1999: from less than 1 million hectares to 28 million in 1998 and approximately 40 million in 1999, and from US$75 million in 1995 to US$1.64 billion in 1998. Private industry has dominated research. Consolidation of the industry has proceeded rapidly since 1996, with more than 25 major acquisitions and alliances worth US$15 billion. Little private-sector agricultural biotechnology research so far has focused on developing country food crops other than maize. Moreover, little adaptation of the research to developing country crops and conditions has occurred through the "enlightened" public and philanthropic channels prominent in the Green Revolution of the developing countries.

Some of the exciting international and regional programmes are described by Cohen. A programme directed at public/private sector linkages is that of the International Service for the Acquisition of Agri-biotech Applications which transfers and delivers appropriate biotechnology applications to developing countries and builds partnerships amongst institutions. Relatively little biotechnology research currently focuses on the productivity and nutrition of poor people. The Rockefeller Foundation's agriculture programme is one example; in 1998, it provided about US$7.4 million for biotechnology research relevant to developing countries, mainly through IARCs and NARS in developing countries, with a major emphasis on rice. This sum pales by comparison with Monsanto's 1998 R&D budget of US$1.3 billion, much of which funded agricultural biotechnology research. As with the Green Revolution, the challenge is to move from the scientific foundation established by industrial country-oriented research efforts to research focused on the needs of poor farmers and consumers in developing countries. Direct transfers of the fruits of agricultural biotechnology research to the developing countries will not work, in most cases. More appropriate research for the developing world might focus on biotechnology and conventional breeding to develop alternative forms of weed resistance, such as leafier rice that denies weeds sunlight rather than incorporating herbicide tolerance into rice.

The West Africa Rice Development Association, a public IARC in Côte d'Ivoire, has used a combination of conventional plant breeding and tissue culture to develop such rice. Insect-resistant crops would have great potential value for poor farmers. So far, however, the development of crops containing genes from the Bacillus thuringiensis bacterium, which produces a natural pesticide, has focused largely on the crops and cropping environments of North America. The new crop varieties containing the Bt gene require extremely knowledge-intensive cultivation. They might well be transferable to larger scale operations in some developing countries such as Argentina. The potential usefulness of this application in crops grown by small farmers is open to question.

There is considerable debate about risks of the development of resistance in pests, harm to beneficial insects, and crosspollination of wild and weedy plants with the novel gene. The evidence on these issues is still inconclusive and warrants careful monitoring before the application of Bt is tried on a large scale in crops grown by subsistence farmers. Research on crops and problems of relevance to small farmers in developing countries will require the allocation of additional public resources to agricultural research, including biotechnology research, that promises large social benefits.

There is no reason to believe that this research will offer lower rates of return than other agricultural research and development. Private-sector agricultural research currently accounts for a small share of agricultural research in most developing countries. The public sector can expand private-sector research for poor people by converting some of the social benefits to private gains, for example, by offering to buy exclusive rights to newly developed technology and make it available either for free or for a nominal charge to small farmers. The private research agency would bear the risks, as it does when developing technology for the market.

IARCs have an important role to play as intermediaries in facilitating such arrangements. Without more enlightened adaptation, continued expansion of genetically improved crop production in the industrial countries may well have a negative impact on small farmers in developing countries. Some developing country consumers would benefit, but those consumers who also farm could experience net losses.

In addition, the development of industrial substitutes for developing country export crops, such as cocoa could have a devastating impact on developing country farmers' livelihoods. In sum, the biggest risk of modern biotechnology for developing countries is that technological development will bypass poor farmers and poor consumers because of a lack of enlightened adaptation.

It is not that biotechnology is irrelevant, but that research needs to focus on the problems of small farmers and poor consumers in developing countries. Private sector research is unlikely to take on such a focus, given the lack of future profits. Without a stronger public sector role, a form of "scientific apartheid" may well develop, in which cutting edge science becomes oriented exclusively towards industrial countries and large-scale farming.

Sessions from the Green Revolution

The outcomes of the Green Revolution offer some guideposts for assessing the likely risks and benefits of agricultural biotechnology for developing countries. Risks and benefits may be inherent in a given technology, or they may transcend the technology. The policy environment into which a technology is introduced is critical. For example, IFPRI research has found that in Tamil Nadu State in India, the adoption of high-yielding grain varieties meant not only increased yields and cheaper, more abundant food for consumers, but income gains for small and largerscale farmers alike, as well as for nonfarm poor rural households. Increased rural incomes contributed to nutrition gains for these households.

Because the Tamil Nadu state government has pursued active poverty alleviation strategies, including extensive social safety net programmes and investment in agriculture, rural development, and a fair measure of equity in access to resources such as land and credit, the benefits were widely shared. Where increased inequality followed the adoption of Green Revolution technology, it was not because of factors inherent to the technology, but rather a result of policies that did not promote equitable access to resources. And even in these areas, rural landless laborers usually found new job opportunities as a consequence of increased agricultural productivity, particularly where appropriate physical infrastructure and markets developed. Successful adoption of Green Revolution technology, however, depended on access to water, fertilizer, and pesticides.

Thus, inequality between well-endowed and resource-poor areas increased because of the properties of the technology itself. Likewise, excessive or improper use of chemical inputs led to adverse environmental impacts in some instances. This problem was offset, to some extent, by characteristics that were also inherent in the technology: by allowing yield gains without expanding cultivated area, the technology kept cultivators from clearing forests and moving onto wild and marginal lands. Overall, the Green Revolution was extremely successful in enhancing productivity in rice, wheat and maize; in increasing incomes and reducing poverty; and in preserving forests and marginal lands by improving yields within existing cultivated areas. By reducing unit costs and prices for food, it greatly benefited poor consumers, and by boosting farmers' incomes, it contributed to gains in nutrition.

Would agricultural biotechnology produce similar results in developing countries? The answer depends on whether the research is relevant to poor people and on its ownership, that is, the nature of the intellectual property rights arrangements.

CAUSES OF FOOD INSECURITY

Food is the most basic of human needs. Despite the "green revolution" between 1970 and 1990 almost half of the world's less developed countries suffer a decline in aggregate food supply, and more than a quarter suffer an increase in hunger. Malnutrition is a major barrier to economic and social development, leaving populations unable to maintain normal lives and to be economically and socially less productive.

POPULATION GROWTH AND URBANIZATION

Within the next 20 years, more poor and under-nourished people in developing countries will live in the cities than in rural areas. High rates of urbanization mean that urban food insecurity and malnutrition are concerns even for regions like Africa and Asia, where current levels of urbanization are relatively low. By the year 2030, the rural population would have grown by more than 235 million, but the urban population would have grown by 2.4 billion. The number of people living in cities in Africa will be more than triple, from 251 million to 864 million.

POVERTY

Food availability means that the overall supply should potentially cover overall nutritional needs in terms of quantity and quality; furthermore, it should be safe and of good food quality. Also, the types of foodstuffs commonly available should be culturally acceptable. Food expenditures can make up as much as 60 to 80 per cent of total income among low-income urban households. The importance of being able to earn cash income also means that the ability to stay healthy, to get a good job and to have access credit to smooth consumption, or expand, or start up businesses, are all critical to urban food and nutrition security. With enough income, prices can rise and families can still buy enough to eat. Millions of urban poor, however, are vulnerable to price rises or sharp declines in income, say due to illness or loss of job by the principal income-earner. The majority of the urban labour force works in sectors like petty trade and services where wages are low and job tenure uncertain. In urban Nigeria and most of sub-Saharan Africa, employment in sectors that pay regular wages, such as manufacturing and industry, accounts for less than 10 per cent of total employment. Urban poverty thus is not primarily the result of lack of work but the lack of well-paying, steady jobs.

HEALTH

Health is determined by a series of factors that act at three levels. At the community level, factors such as the quality of the overall environment and the availability, cost, and quality of services such as water, electricity, sewage, refuse disposal, and health services are important health determinants. At the household level, the most important factors include the general conditions of the household, including the type of housing, the availability and cost of water and hygienic facilities, and the number of rooms per household member;

the availability of food; and household caring behaviours related to the use of preventive and curative health services, the use of water and hygienic facilities to provide a healthy, hygienic and safe environment, and food-related behaviours such as the acquisition of food, the intra-household allocation of resources, feeding practices and food preparation methods.

At the individual level, the determinants of health relate to the interactive mechanisms among an individual's food and nutrient intake, nutritional status, and health status. High rates of malnutrition and escalating rates of dietrelated diseases such as diabetes and hypertension in developing countries are attributable to various concerns: inadequate sources of dietary protein, foods with high levels of anti-nutritional components and toxicants, a disproportionate amount of highly digestible, high glycaemic index carbohydrates which constitute the staple foods such as yams, maize and rice, and limited alternatives.

In addition, socio-economic factors such as urbanization and migration to urban areas have led to changes in lifestyles to include imported highly processed foods and modification in eating patterns and food habits. In many rural areas, there are challenges of protein malnutrition as well as inadequate vitamin and mineral intake.

Furthermore, these problems are compounded by inadequate prenatal nutrition which leads to deficiencies in both mothers and children. Nutritional needs of postpartum nursing mothers are often not met, subsequently leading to inadequate nutrition in children.

The quality of staple foods and other foods that are commonly consumed in most of these areas could be optimized to improve their quality. In developing countries where mechanization is still limited, human labour provides much of the power for productive economic work. For the poor, labour is their most important asset. Work capacity, performance, and productivity of workers are therefore generally important for income. This can lead to a situation where poor, malnourished workers perform poorly because of decreased working capacity, and therefore do not get access to better-paying jobs. Various studies support this assumption, showing positive associations between wage achievement and nutritional status as proxied by either body weight, body-mass index, stature, or caloric and protein intake. Furthermore, the poor sanitary conditions of many developing-countries are notorious, and this combined with poor hygiene practices, create a situation in which food safety is severely compromised.

OUTDATED FARMING SYSTEM

Small-scale farmers in developing countries are faced with many problems and constraints. Pre- and postharvest crop losses due to insects, diseases, weeds, and droughts result in low and fluctuating yields, as well as risks and fluctuations in incomes and food availability. Low soil fertility and lack of access to reasonably priced plant nutrients, along with acid, salinated, and waterlogged soils and other abiotic factors, contribute to low yields, production risks, and degradation of natural resources as poor farmers try to eke out a living.

They are often forced to clear forest or farm ever more marginal land to cultivate crops. Poor infrastructure and poorly functioning markets for inputs and outputs together with lack of access to credit and technical assistance add to the impediments facing these farmers.

POLITICS

Eight hundred million people on earth are poor and malnourished. They live on less than a dollar a day and cannot be sure that their fields will yield enough food or that they will earn enough money to buy food. Forty thousand people die each day of malnutrition, one-half of them children. The increase in food production enabled by the Green Revolution unfortunately did not solve the problems of malnutrition and hunger.

There were about a billion hungry people some 40 years ago, and population projections show that there may still be 600 million poor people by 2025, when the earth's population would have grown to 8 billion. The Green Revolution did many things, but it did not wipe out poverty. Not enough jobs were created in either the rural areas or the cities to generate the purchasing power that provides farmers with the incentive to grow more food.

It is ironic that hunger persists while the prices for agricultural commodities are at an all-time low. Several countries that have millions of hungry people are exporting food and other agricultural products to countries where people are already well fed. Yet, most of these countries that are poor, with so many hungry people, seem to be able to grow food quite abundantly.

3

A Potential Role for Modern Biotechnology in Agriculture

The Convention on Biological Diversity dictates the use and application of relevant technologies as a means of achieving the objectives of conservation and sustainable use with specific reference to biotechnology. Modern biotechnology is purported, from a technical perspective, to have a number of products for addressing certain food-security problems of developing countries. It offers the possibility of an agricultural system that is more reliant on biological processes rather than chemical applications.

The potential uses of modern biotechnology in agriculture include: increasing yields while reducing inputs of fertilizers, herbicides and insecticides; conferring drought or salt tolerance on crop plants; increasing shelf-life; reducing postharvest losses; increasing the nutrient content of produce; and delivering vaccines. The availability of such products could not only have an important role in reducing hunger and increasing food security, but also have the potential to address some of the health problems of the developing world. Achieving the improvements in crop yields expected in developing countries can help to alleviate poverty: directly by increasing the household incomes of small farmers who adopt these technologies; and indirectly, through spill-overs, as evidenced in the price slumps of herbicides and insecticides. Indirect benefits as a whole tend to have an impact on both technology adopters and non-adopters, the rural and urban poor. Indeed, some developing countries have identified priority areas such as tolerances to alkaline earth metals, drought and soil salinity, disease resistance, crop yields and nutritionally enhanced crops.

The adoption of technologies designed to prolong shelf-life could be valuable in helping to reduce postharvest losses in regionally important crops. Prime candidates in terms of crops of choice for development are the so-called 'orphan crops', such as cassava, sweet potato, millet, sorghum and yam. Multinationals have found no incentive to develop these crops and have instead invested in marketable crops with high profit returns. This strategy is intended to target wealthier farmers in temperate-zone countries with the financial capacity and tradition of supporting new seed products. However, here is a potential for multinational companies to develop crops grown largely in developing countries. The investment costs are low and the potential markets considerably large. While some public-sector research institutes in developing countries are forging ahead with the application of modern biotechnology, a small number are supported by government policy and therefore follow a defined agenda.

Still other governments believe that the risks associated with modern biotechnology outweigh the benefits. Currently, the many promises of modern biotechnology that could have an impact on food security have not been realised in most developing countries. In fact, the uptake of modern biotechnology has been remarkably low owing to the number of factors that underpin food security issues. In part, this could be because the first generation of commercially available crops using modern biotechnology were modified with single genes to impart agronomic properties with traits for pest and weed control, and not complex characteristics that would modify the growth of crops in harsh conditions. Secondly, the technologies are developed by companies in industrialized countries with little or no direct investment in, and which derive little economic benefit from, developing countries. Thirdly, many developing countries do not have the necessary biosafety frameworks to regulate the products of modern biotechnology.

For example, it took over two years for the Kenyan authorities to approve the field-testing of a virus-resistant sweet-potato variety because the scientific capacity for evaluating the product was unavailable. It should be noted, however, that such delays in the approval process have also been seen in developed countries, especially during the initiation of national regulatory evaluation. However, this trend is quickly changing as a number of developing countries either adopt or develop appropriate biotechnologies or regulatory infrastructures.

A report by the International Service for National Agricultural Research states that more than 40 crops are the focus of public-sector research programmes of 15 developing countries involving disease-resistant traits in rice, potato, maize, soybean, tomato, banana, papaya, sugarcane, alfalfa and plantain. For example, the Brazilian Agricultural Research Corporation has concentrated research into genetically modified crops on disease resistance in beans, papaya and potatoes.

The research programme at the University of Cape Town focuses on the development of crops resistant to viruses and to desiccation. The university has recently had a breakthrough with maize streak-virus resistance. In Thailand, the National Centre for Genetic Engineering and Biotechnology has supported research into disease resistance of rice, pepper and yard-long beans. therefore follow a defined agenda. Still other governments believe that the risks associated with modern biotechnology outweigh the benefits. Currently, the many promises of modern biotechnology that could have an impact on food security have not been realised in most developing countries. In fact, the uptake of modern biotechnology has been remarkably low owing to the number of factors that underpin food security issues.

In part, this could be because the first generation of commercially available crops using modern biotechnology were modified with single genes to impart agronomic properties with traits for pest and weed control, and not complex characteristics that would modify the growth of crops in harsh conditions. Secondly, the technologies are developed by companies in industrialized countries with little or no direct investment in, and which derive little economic benefit from, developing countries. Thirdly, many developing countries do not have the necessary biosafety frameworks to regulate the products of modern biotechnology.

For example, it took over two years for the Kenyan authorities to approve the field-testing of a virus-resistant sweet-potato variety because the scientific capacity for evaluating the product was unavailable. It should be noted, however, that such delays in the approval process have also been seen in developed countries, especially during the initiation of national regulatory evaluation. However, this trend is quickly changing as a number of developing countries either adopt or develop appropriate biotechnologies or regulatory infrastructures.

A report by the International Service for National Agricultural Research states that more than 40 crops are the focus of public-sector research programmes of 15 developing countries involving disease-resistant traits in rice, potato, maize, soybean, tomato, banana, papaya, sugarcane, alfalfa and plantain. For example, the Brazilian Agricultural Research Corporation has concentrated research into genetically modified crops on disease resistance in beans, papaya and potatoes. The research programme at the University of Cape Town focuses on the development of crops resistant to viruses and to desiccation. The university has recently had a breakthrough with maize streak-virus resistance. In Thailand, the National Centre for Genetic Engineering and Biotechnology has supported research into disease resistance of rice, pepper and yard-long beans.

Developing countries with limited financial and human resources need to find the right balance for investing in conventional and modern biotechnology research programmes. While alliances with the private sector may contribute to the search for new technologies, the public sector needs to focus on crops and traits in which the former may be unwilling or unable to invest.

The extent to which priority is given to modern biotechnology over other research methods should be linked to a country's agricultural priorities and objectives as well as to its environmental concerns. Ultimately, investment in interventions that support good governance, the development of rural infrastructure and market access is required before any of the promises of modern biotechnology can be realised. In general, policies that stimulate economic growth and target poverty reduction may have significant bearing on the health and well-being of the population.

RESEARCH OWNERSHIP

Research is a critical part of any effort aimed at improving food production and reducing poverty. Globally, much of agricultural R&D is carried out in the public sector, thereby serving the interests of developing countries. Public research in developed countries and Latin America is mostly conducted by government institutions and universities, whereas almost all agricultural research in Africa is carried out in public institutions, including R&D, technology transfer and dissemination of improved plant varieties. In general, international agricultural research institutions form a second level of research development and technology providers in developing countries. Public research institutes have, in the past, researched and improved orphan crops, mainly for donation to poor farmers or at cost. Generally, academic institutions are perceived as producers of knowledge that benefit and protect the public. Also, national and international research institutes aim to address the agricultural problems of resource-poor farmers in developing countries, *e.g.* increasing productivity through the use of a variety of techniques, including modern biotechnology. In reality, public institutions are now exposed to the forces of globalization and compelled to compete for their survival. In the current climate, government intervention in R&D worldwide has dwindled; hampering the level of innovation generated for public good. In fact, research facilities in many developing countries are poorly equipped, often limiting experiments to traditional and outdated research.

The diminishing role of public research institutes is perceived to have a major impact on the adoption of modern biotechnology in terms of introducing relevant products to those that need them most. Most of the field trials in the EU and the USA are conducted by private companies. An analysis of field-trial data from the USA shows that three crops account for 64% of all trials, of which 69% express herbicide- and pest-resistance traits. Of the trials conducted in the EU, 67% involve maize, sugar beet and rapeseed, and 71% of the novel gene categories presented herbicide- or pest-resistance traits. Less than 1% of all the trials in the EU and the USA are of plant varieties grown in tropical and subtropical climates, half of which have been conducted by the public sector.

Most of the public-sector research involving modern biotechnology in developing countries is still in the laboratory phase - none of the crops have progressed to marketable products. China, on the other hand, has approved the field-testing of over 500 GMOs to date and the commercial release of 50, including a prolonged shelf-life tomato, virus-resistant sweet pepper and vaccines for animal use. The experiences limiting the progression to commercializing research efforts range from: a lack of resources for meeting the high costs of regulatory requirements; lack of foresight, planning and business acumen for enabling the transition from research to a commercial product; and lack of capacity to negotiate patent licenses. Also, developments in modern biotechnology have occurred independently of the sustainable agricultural goals and priorities of the developing countries concerned. Furthermore, a needs assessment for a particular technology has often not been carried out before a research project is begun. Nevertheless, it is often argued that the commercialization of some products would encourage monocropping as national agricultural research has focused on a few crops, whereas rural communities tend to grow a wide range of crop species and plant varieties. The focus of international agricultural research centres is on plant production and protection, livestock production and health and food processing. With respect to food crops, research emphasis appears to be spread equally between cereals, root crops and legumes. However, within the cereals, research devoted to rice far outweighs research on maize and sorghum.

A large proportion of the total agricultural research activities of many developing countries are donor-funded. International research institutes, such as the Consultative Group on International Agricultural Research, depend on government grants and donations from philanthropic organizations for their survival, and yet investment in this sector has fallen in real terms. A significant fraction of the funding spent on international agricultural research institutes is used on activities covering a relatively large number of crops. The beneficiaries of such initiatives are a small number of countries with relatively advanced scientific capabilities. During the 1990s, developing countries as a group invested more in agricultural research than developed countries, even though the spending was unevenly distributed. In industrialized countries, private-sector investment in R&D far exceeds government spending on technology development, so that much of the public good previously entrusted to public research institutes is now privately owned. In comparison, private-sector investment in developing countries is around 1% of total global spending in this sector and developing countries invest less than 5% of the total private sector spending in biotechnology. Although the agricultural sector in developing countries is large and of significant importance to the domestic economy, spending in agricultural research does not match this level of activity. For instance, 80% of the food consumed in sub-Saharan Africa is obtained from domestic production.

Impact of Intellectual Property Rights on Research

Intellectual property rights have been relevant to agriculture since their inception but have gained importance with regard to research in developed countries in the past 20-30 years. In particular, IPRs have been used to protect and preserve the value of products produced by conventional methods, *e.g.* the trademark registration of food products. The rationale for IPRs is that they encourage the inventor to advertise the invention and disclose the new knowledge, while simultaneously holding the rights to protect the invention from competitors. Disseminating this information is thought to stimulate new ideas and further rounds of innovation and technological advancement. IPRs afford time-limited protection to artistic, scientific, technological or economic products, and can be protected by way of copyrights, trademarks, design patents, utility patents, plant patents, plant breeders' rights and trade-secret law. Of these mechanisms, patents are considered the most powerful tool of the IPR system. Patents play different roles in different technologies and sectors. Patent protection of biotechnology makes it a tool for technology transfer and securing new markets in a global economy. Without protection, new ideas and information are entirely in the public domain.

This can, in certain systems, lead to underinvestment in R&D or the withholding of knowledge. Plant variety protection provides less protection than patents in that generally it makes provision for farmers' rights, allowing them to use harvested seed, and includes an exemption for research use. Despite the increase in availability, new plant varieties continue to be inaccessible or inappropriate for poor farmers, and the rate of innovation remains largely unchanged in countries with a PVP system. Studies have indeed shown that in middle-income countries, the principal beneficiaries of PVPs are commercial farmers and the seed industry. PVP is seen as a system that protects small advances in plant breeding, while a patent regime is thought to lead to the protection of big leaps in technological achievements. Patent protection for products of modern biotechnology is important because they are expensive to develop and easy to copy. Even so, developing countries have limited capabilities to innovate in industrial fields such as modern biotechnology, and to effectively enforce IPRs. A significant number of developing countries have not established intellectual property regimes that cover plants.

This situation may thus discourage private-sector investment. With no assurance that they can recoup some profit on GM products, multinationals are unlikely to devote much attention to the challenges of developing countries unless seen in a development aid context or through public-private partnership. Although this situation impedes private-sector investment in developing countries, it also implies that the freedom to operate on products designated for local markets is not hindered. Exercising this freedom to operate is not a well-understood concept. For example, in the case of 'golden rice', where permission was required for the use of about 70 patents, the impression was that the patents were being relinquished in favour of the poor. In fact, most of the patents involved are not valid in the major rice-consuming countries.

The technology donated for the development of virus-resistance in non-commercial potato varieties is free of patents relevant to Mexico, and the same holds true in the case of virus-resistant sweet potatoes in Kenya. Researchers are usually unaware of the proprietary status of the technologies they are using in their work. Nevertheless, the proliferation of broad patents is thought to impede the research capabilities of other interested parties. Some countries grant very broad patents conferring monopoly rights over large areas of research, thereby potentially threatening the other goal of intellectual property, namely the right to build upon the original invention. The prevailing patent rules have the potential to limit the accessibility of these technologies to public institutions and ultimately poor farmers.

Furthermore, the strengthening of IPRs is thought to restrict the flow of germplasm and inhibit the development of new plant varieties. This is because if and when researchers in public institutions do get permission to develop the technologies further, access is granted under licence agreements with restrictions on commercializing innovations. It is also argued that a stringent, multilateral IPR system will not benefit all countries equally. Indeed, the benefits will largely be influenced by the economic and technological levels of development in each country.

The United Kingdom Commission on Intellectual Property Rights, "the critical issue in respect of IPRs is perhaps not whether it promotes trade or foreign investment, but how it helps or hinders developing countries to gain access to technologies that are required for their development." During the 1990s, obtaining a patent application in the USA cost US$20,000 and twice as much in the EU. In general, PVP is cheaper; valued at one-tenth of the patent price. Moreover, the preparation of a food-safety dossier for a product derived from modern biotechnology, for example, is estimated to be around US$1 million.

These estimates cannot be compared with the regulatory costs in developing countries. The cost of regulation in developing countries does not encourage the commercialization of products of modern biotechnology developed by public-sector research institutes. In most cases, the regulatory costs far exceed the research costs. Many developing countries do not have the resources to match private-sector investment in modern biotechnology. In this new playing field, public institutions also need resources to deal with intellectual property rights to help compensate and increase the public benefit. Otherwise their involvement in R&D could be deterred by lack of funding. If public institutions are to use the techniques of modern biotechnology, then the use of IPRs as a framework for facilitating technology transfer must be emphasized more than its handling as a revenue-generating system. IPRs can, however, play a major role in clarifying the mechanisms for access to technology, and determine the downstream aspects of use and exploitation of genetic resources.

There are several ways in which public institutions and small companies in developing countries can gain access to patented genes and enabling technologies to overcome the current barriers to research. The first of these includes a measure of goodwill by multinationals to relinquish their rights to technologies for use by researchers in developing countries by adopting programmes of social responsibility as in the case of 'golden rice', a variety containing beta-carotene, virus-resistant sweet potatoes and a virus-resistant non-commercial potato variety in Mexico.

A different type of programme initiated in the USA has established an intellectual property clearing house to make information on the intellectual property owned by public research institutes, including universities, available to researchers around the world.

There are also suggestions that redesigning patent laws to narrow the type and scope of patent coverage ought to make more technologies accessible to public institutions. The thinking behind these suggestions is that applying a stronger standard for rejecting patent applications for inventions that are 'obvious' should deter the patenting of minor inventions. In addition, a law that requires an invention to be genuinely useful in theory should reduce the number of patent applications being submitted. At present, it is possible in some countries to submit patent applications for abstract concepts that potentially protect large areas of research and thereby exclude innovation by others. Another option that may be attractive to developing countries is the creation of collaborations that involve research institutes, universities and the private sector.

The nature of these collaborations is likely to be influenced by the level of expertise and resources within the national public research institutes. Where a solid knowledge base exists, the public partners may be in a position to develop or acquire a technology that could be transferred into locally adapted varieties. Smaller institutes are more likely to provide the genetic resources and a positive public image. It is believed that such alliances would benefit public institutions and private companies; offering them an opportunity to license and distribute the technology.

The most well-known public-private sector partnerships include organizations such as the International Service for the Acquisition of Agri-biotech Applications that negotiate access to private-sector technologies for the improvement of subsistence crops and/or the transfer of technology and know-how. Although several types of public-private sector alliances already exist, the two newly established initiatives worth mentioning are the Global Cassava Partnership and the African Agricultural Technology Foundation, launched in November 2002 by FAO and in March 2003 by the Rockefeller Foundation, respectively. The former is a partnership involving some of the world's leading experts in cassava research, working mainly in public institutions. The AATF intends to function as a clearing house of available technologies with the primary aim of improving food security and reducing the poverty situation of small farmers, by facilitating transfer and use of appropriate technologies.

Such licensing arrangements have been put to the test in other fields. However, as in the case of the AATF, a clearing house is required to acquire the necessary technologies and permit their further use for developing-country needs. The drawback may be a requirement to divide the commercial sector into subsistence, middle-income and commercial markets. This market division may not be easy to achieve as some large developing countries have both commercially important markets and subsistence farmers.

A briefing document commissioned by the United Nations Industrial Development Organization proposes six activities that build upon private-sector investment and enable biotechnology transfer:

- Enabling government policies;
- Access to up-to-date authoritative information;
- Regional brokering service to strengthen public-private partnerships;
- A regional biotechnology investment service;
- An international intellectual property escrow service; and
- Initiatives for risk-shifting.

Each of the proposals can be implemented as a stand-alone project or in combination, as best suits the national and/or regional situation.

Access to Genetic Resources

Historically, plant genetic resources were freely provided by developing countries to gene banks worldwide. The resources in question did not belong to a particular individual, and are often considered a common heritage of mankind. The application of modern biotechnology to genes that could be incorporated in genetic resources of importance in rural populations raises concerns in that the small-scale farmers may have originally supplied the genetic resources for improvement.

Once privately owned, these resources may be unavailable to the people who have ensured their conservation for centuries.

Equally important is the aspect of access by researchers to genetic resources for further development on terms that recognize the contributions made by farmers to the conservation and sustainable utilization of these resources. At the international level, the importance of national ownership of such resources is duly recognized.

The International Treaty on Plant Genetic Resources adopted at an FAO conference in 2001 provides the legal framework for dealing with the resources on which food security and sustainable agriculture depend. The Treaty gives a directive on the conservation and sustainable use of plant genetic resources for food and agriculture, making provision for the fair and equitable sharing of the benefits arising out of their use, in harmony with the principles of the

Convention on Biological Diversity, but introducing the concept of farmers' rights. In the discussions on farmers' rights, the main issues of concern revolve around benefit-sharing and prior informed consent, and the protection of traditional knowledge from 'biopiracy'. This means access to genetic resources must be on mutually agreed terms to promote their use and emphasize their importance to development. A number of organizations are discussing the protection of traditional knowledge and folklore.

The Treaty establishes a multilateral system of facilitated access and benefit-sharing for key crops, emphasizing the interdependency of countries in terms of plant genetic resources for food and agriculture. Developing countries rich in genetic resources are encouraged to place germplasm in the MLS. The users of the material will sign a material transfer agreement, incorporating the conditions for access and benefit-sharing through a fund established under the Treaty. In return, the owners of the genetic resources will get a share of the benefits arising from their use and development in the way of information, technology transfer and capacity building. *Ex situ* material, collected before the entry into force of the CBD, does not come under the purview of the Convention, and would thus be dealt with under the Treaty.

So far, 35 food and 29 feed crops have been entered into the system. In principle, the genetic resources stored under this system are available for improvement to all interested researchers. Such wide-scale availability of germplasm has a potentially positive influence on access to improved technologies and nutritionally enhanced staple crops for the food-insecure. Genetic resources obtained from the MLS cannot be patented, even though it is not clear whether a gene isolated from such material may be protected or not.

GLOBALIZATION

Globalization is a complex set of developments including trade liberalization, opening up of economies and integration of international markets. It involves policy reforms in trade and reducing barriers to the international flow of goods, capital, labour, technology and ideas. Technological advancement is the driving force of improvements in world food production and the integration of the world economy.

Knowledge in agricultural science depends on building onto experiences already gained. In this respect, the introduction of biotechnology ought to be perceived as nothing but a phase in the modernization of agriculture which started centuries ago. In its evolution, a global trend is emerging where fewer farmers produce more food.

The global commercialization of agriculture has increased competition in domestic and international markets. Globalization has accelerated substantially since the 1980s, as the economies of developing countries have become subject to international market forces. New technologies appear to drive the transformation of world food systems towards industrialized food processing, long-distance marketing and retail-business dominance. The benefits of globalization, however, seem to have bypassed low-income countries and, in particular, rural regions. Even though the profit potentials were different, the merging of agrochemical and pharmaceutical divisions to form life sciences companies in the mid-1980s was a strategic synergy of R&D that allowed the production of new drugs, pesticides, GM crops and genetic treatments for disease. This has led to new patterns of alliances with companies that develop routes to create and capture value. The integration of economies cuts across all sectors and likewise is central to the emergence of new patterns of R&D. The phenomenon of globalization has raised concerns about the global acquisition of seed companies and mergers with chemical interests that have strategically strengthened their ability to market new products, and placed a substantial number of agricultural patents in the control of five major companies, namely Bayer, Dow Chemicals, Du Pont, Monsanto and Syngenta.

The top 10 seed companies control one-third of the global seed market. This kind of consolidation has also placed much of the know-how under the control of these multinationals. To overcome problems of overlapping patent rights and reciprocal infringement lawsuits, companies make strategic alliances, mergers or acquisitions to obtain technologies in specified fields of research. Similar consolidation trends have been observed in developing countries, where PVP legislation is the only means of protecting improved seed—resulting in a proportionately small number of companies dominating global markets related to food and agriculture. Overall, the seed industry can be divided into three segments: commercial seed, farm-saved seed, and public-supplied seed. Farmers' sources of seeds tend to be flexible in that they respond to local needs and circumstances and therefore can greatly vary for the same crop just as to location. Farmers in most developing countries depend on farm-saved and public-supplied seeds.

The latter provide seeds for the most important crops, while farm-saved seeds account for over 90% of the planted crops. Global consolidation of seed companies is likely to homogenize seed quality and limit choice. By the same token, the same genetic traits can be introduced into locally adapted varieties in different world regions. It is important, then, that when considering the effects of consolidation in the commercial seed market, it should be kept in mind that the commercial seed market is just one segment of the total seed market and, since it tends to concentrate on high-value seeds, provides less than 20% of the planted seed material in both developed and developing countries. Furthermore, it should be recognized that GM seed represents only a minority of the commercial seed market, *i.e.* approximately 13%.

MARKET ACCESS

More than 50% of the workforce in developing countries makes a living from agriculture, and so it follows that economic development in these countries relies on agricultural performance. The most important indicators of agricultural performance are an increase in domestic production and accelerated access to markets. Trade liberalization, as negotiated in the Uruguay Round, was intended to create opportunities that improve market access for the commodities of developing countries by eliminating agricultural subsidies, high import duties and other trade-distorting measures adopted by developed countries.

At present, agriculture is subsidized at US$1 billion per day in the OECD countries, making it impossible for poor countries to compete on an equitable basis. Limited market access represents the greatest hurdle to international trade, and consequently to technology access and acceptance. If developing countries are to be integrated into the global economy, they must have something to offer that in turn translates into beneficial implications for food security. Policy options for those developing countries whose exports largely rely on agricultural trade will be influenced by the regulatory climate and consumer preferences in trading-partner countries, including agricultural-subsidy schemes that make the products of developing countries less competitive.

The continued support by developed countries of high agricultural subsidies creates inequitable trading conditions for developing countries. In addition to the two major barriers to trade for developing countries, a third hurdle in relation to the products of modern biotechnology may be the implementation of the CPB. The CPB adopted in 2000, entered into force in September 2003. A legally binding instrument to its parties, it aims to regulate trade in products of modern biotechnology by protecting biodiversity and taking also into account effects on human health. The backbone of the CPB is an advanced informed agreement that requires the party of export to seek consent from the party of import before the first shipment of living modified organisms intended for release into the environment. A simplified procedure is mandated for commodity trading. This is based on the proactive information exchange between the party of export and its potential trading partners via an Internet-based Biosafety Clearing House. The process is intended to facilitate trade by providing easy access to data and thereby enabling early assessments in prospective recipient countries.

This, in effect, makes available information on the regulatory status of a new trait in any country before the initiation of trade or the delivery of food aid. Timely handling of this information by developing countries will be put to the test now that the CPB has entered into force. The SPS Agreement of WTO ensures that internationally traded food meets minimum standards based on the scientific principles set by the Codex Alimentarius Commission. It is expected that when dealing with commodities containing products of modern biotechnology, the SPS Agreement will make reference to the Codex

Principles for the risk analysis of foods derived from modern biotechnology and the guidelines for risk/safety assessment of such foods. The Codex Alimentarius Commission approved the principles and guidelines in 2003. The SPS Agreement both offers opportunities for improving food safety in developing countries and poses potential problems.

The science and research-based requirements of the SPS Agreement can be substantial. Yet for countries that can implement the Codex standards for their export products, the opportunity to expand trade and acquire much-needed foreign currency is guaranteed. Those developing countries that cannot find resources to improve their food safety systems may find themselves excluded from international trade. More importantly, the limited capacity to do scientific risk assessment can compromise their ability to assess whether or not to import the products of modern biotechnology. The increased international trade in food has created greater dietary diversity, year-round availability and often low prices in many countries. Non-traditional foods are becoming increasingly available worldwide.

Producers in developing countries can benefit from expanded food exports which earn foreign exchange and increase rural incomes. These benefits, however, may not be realised if food safety and quality standards required in high-income markets cannot be met. Capturing these new opportunities places the onus on developing countries to manage safety from farm to table. Testing of hazards at several points in the production process is expensive, and therefore prevention and control through documented production practices is often the only way to verify food safety. The concept of process control and hazard prevention is generally well understood in developed countries.

When a product is consumed domestically, and investments to meet export market standards for that product affect a large portion of production, those investments will have positive spill-over for domestic consumers. However, some products may be produced almost entirely for export, in which case investment to meet high standards of safety and quality in export will have little or no direct spill-over for domestic safety. Nielsen and Anderson looked into the policy ramifications resulting from the economic effects of adopting GMOs.

The study highlights the implications for the global economy when selected regions adopt modern biotechnology, in terms of agricultural production, trade and economic welfare. Analysis is made of the trade in cereal grains and oilseeds.

The authors suggest that similar analyses of crops unlikely to cause food-safety concerns in other areas and that are of potential economic importance to developing countries should be conducted. They believe prejudices in the more affluent countries may hinder the adoption and production of all products of modern biotechnology in all regions, unless more informed debates are held to discuss the potential opportunities for developing countries. The United Kingdom Government's Strategy Unit also concluded that trade, policies and consumer behaviour in the EU and United Kingdom would influence the decisions of developing countries related to the use of GM crops. While export markets may be important for the rural poor, meeting higher standards may require additional management, capital investment, purchased inputs, monitoring and certification. Government focus on public-health issues will not necessarily address the export barriers.

APPLICATIONS OF MODERN BIOTECHNOLOGY IN FOOD CROPS

In 1990, the U.S.Food and Drug Administration approved the first genetically engineered food ingredient for human consumption, the enzyme chymosin, used in cheesemaking. It is estimated that today 70% or more of cheese made in the U.S. uses genetically engineered chymosin. The first genetically engineered food, the FlavrSavr™ tomato, was approved for human consumption in the U.S. in 1994.

Transgenic Acreage Expands Steadily

Seven million farmers in 18 countries now grow genetically engineered crops. Leading countries are the U.S., Argentina, Canada, Brazil, China, and South Africa. Cultivation of genetically engineered crops globally has expanded more than 10% per year for the past seven years, just as to the International Service for the Acquisition of Agri-biotech Applications. Such an expansion rate amounts to a 40-fold increase in the global area of transgenic crops from 1996 to 2003. Thus, in spite of continuing controversy, the technology continues to be adopted by farmers worldwide.

ISAAA highlighted its key findings this way:

- In 2003, GM crops were grown in 18 countries with a combined population of 3.4 billion, living on six continents in the North and the South: Asia, Africa and Latin America, and North America, Europe and Oceania.... the absolute growth in GM crop area between 2002 and 2003 was almost the same in developing countries and industrial countries ... the three most populous countries in Asia-China, India, and Indonesia, the three major economies of Latin America-Argentina, Brazil and Mexico, and the largest economy in Africa, South Africa, are all officially growing genetically engineered crops.

The leading genetically engineered crops globally and in the U.S. are soy, maize, cotton, and canola. In the U.S., transgenic virus-resistant papaya and squash are also cultivated.

Agronomic Traits Prevail

Research in plant biotechnology has focused primarily on agronomic traits-characteristics that improve resistance to pests, reduce the need for pesticides, and increase the ability of the plant to survive adverse growing conditions such as drought, soil salinity, and cold. Biotechnology traits developed and commercialized to date have largely focused on pest control or herbicide resistance. Many plant pests have proven either difficult or uneconomical to control with chemical treatment, traditional breeding, or other agricultural technologies and in these instances in particular, biotechnology has proven to be an effective agronomic tool. Herbicide resistance allows farmers to control weeds with chemicals that would otherwise damage the crop itself. Varieties combining two different traits, such as herbicide tolerance and insect resistance, have been introduced in cotton and corn.

The addition of new traits, such as resistance to rootworm in maize, and the combinations of traits with similar functions, such as two genes for resistance to lepidopteran pests in maize, are expected to increase. In its 2003 report, ISAAA suggested that five new Bt and novel gene products for insect resistance in maize could be introduced. While the improvement of agronomic characteristics in major crops has been highly successful, few products genetically engineered to meet the specific needs of either food processors or consumers have yet been commercialized. Recently, however, a renewed emphasis on developing agricultural biotechnology applications more relevant to consumers has accompanied continuing efforts to develop crops with improved agronomic traits. Although genetically engineered crops with enhanced health, nutrition, functional, and consumer benefits have lagged behind agronomic applications, research on many such products is in the advanced stages of development. These applications could improve human and livestock nutrition and health, the nutritional quality of food animals for human consumption, and create ingredients with superior properties for food manufacturing and processing.

FOOD APPLICATIONS FOR HUMAN HEALTH

Quantity and Quality of Food Oils

Food oils have both nutritional and functional qualities. From a nutritional perspective, fats and oils contribute more energy than any other nutrient category, about nine calories per gram. This compares with about four calories per gram from carbohydrates and protein. At the same time, specific fatty acids that comprise most of what we call "fat" can affect a person's risk of developing certain chronic diseases such as heart disease. Research over the past several decades has shown that some categories of fatty acids, such as saturated fatty acids, increase the risk of heart disease and other chronic diseases when consumed in excess. Fatty acids also influence how foods behave during manufacturing and processing. For example, saturated fatty acids add stability, texture, and flavour to foods, so they are not simple to replace. To reduce the saturated fatty acid content of foods, plant breeders and food manufacturers increased their use of vegetable oils rich in polyunsaturated fatty acids and developed food oils low in saturated fatty acids. One example is canola oil with 6% to 7% total saturated fatty acids. To improve the stability of vegetable oils rich in polyunsaturated fatty acids, food manufacturers developed partially hydrogenated oils.

The process of hydrogenation reduced the polyunsaturated fatty acid content and increased oil stability, but created *trans* fatty acids, which were subsequently associated with adverse health effects. As a result, hydrogenated fats, the main source of dietary *trans* fatty acids, are now being eliminated from foods. Food manufacturers are developing other ways to reduce undesirable saturated fat content while maintaining stability such as using short chain saturated fatty acids and monounsaturated fatty acids. To date, one functional food oil created with the tools of biotechnology has been commercialized. Calgene's high lauric acid canola, Laurical™, containing 38% lauric acid, is used in confectionary products, chocolate, and non-food items such as shampoo. Conventional canola oil does not contain lauric acid. Laurical™ is a substitute for coconut and palm oils. FDA approved its use in foods in 1995. The following part describes research to date focused on developing crop varieties with other unique oil profiles.

Strategic Aims of Altered Fatty Acid Profile

Improving the healthfulness and functionality of food oils can be accomplished in several ways.

Where traditional plant breeding reaches its limits, biotechnology may be used to:

- Reduce saturated fatty acid content for "heart-healthy" oils
- Increase saturated fatty acids for greater stability in processing and frying
- Increase oleic acid in food oils for food manufacturing
- Reduce alpha-linolenic acid for improved stability in food processing
- Introduce various omega-3 polyunsaturated fatty acids including long-chain forms
- Enhance the availability of novel fatty acids, *e.g.*, gamma-linoleic acid

QUANTITY AND QUALITY OF PLANT PROTEIN

Efforts to improve the protein content and quality of staple foods have been underway for decades. The main focus is crops grown in developing countries, where nutrient shortfalls are widespread and dietary diversity limited. Foods such as potato and cassava, staple foods in several parts of South America and Africa, have less than one per cent protein. Efforts to improve protein quality strive to increase the amount of limiting essential amino acids provided by the protein in the food. The amino acids most often present in inadequate amounts are lysine, tryptophan, and methionine. Improvements in protein quality benefit both human and animal nutrition and increase the feed efficiency of crops fed to food animals. For example, corn is widely fed to cattle but it is limiting in lysine and methionine. Corn with higher levels of these amino acids would significantly improve feed efficiency and lower input costs to farmers. Improved corn varieties consumed by humans would also have nutritional benefits.

There are various ways of improving protein quantity and quality. One is to increase the total amount of protein produced by selecting germplasm with an altered balance of seed proteins. This may be done by traditional cross breeding or genetic engineering. Another approach is to introduce genes from other sources for proteins that have a favourable balance of essential amino acids. An example is the introduction into potato of a gene for seed albumin protein from amaranth. A third approach seeks to increase the production of specific amino acids such as lysine.

This approach was used in the development of Quality Protein Maize. William Folk and his team at the University of Missouri, Columbia, MO, pioneered another approach to improve seed protein quality. Their strategy was to substitute more desirable and scarce amino acids for more abundant ones in certain seed proteins. They applied this concept to rice by increasing the production of lysine, an essential amino acid, at the expense of the non-essential amino acids, glutamine, asparagine and glutamic acid. *Cassava*: A staple food for some 500 million people in tropical and sub-tropical parts of the world, cassava, also known as yucca or manioc, thrives in marginal lands having little rain and nutrient-poor soils. It is widely consumed in Africa, and parts of Asia and South America. Cassava root has less than 1% protein and poor nutritional value.

However, the leaves are also consumed and these are a good source of beta-carotene, the precursor of vitamin A. In 2003, Zhang and colleagues reported using a synthetic gene to increase the protein content in cassava. The gene is for a storage protein rich in nutritionally essential amino acids. When the gene was expressed in cassava, transformed plants expressed the gene in roots and leaves, both of which are consumed in human diets.

The experiment demonstrated the feasibility of increasing the quantity and quality of protein in cassava. Cassava also contains cyanogenic glucosides that can produce chronic toxicity if not eliminated or reduced by grating, sun-drying, or fermenting. Efforts to develop cassava varieties low in these toxicants is a high research priority. *Corn*: Corn is the predominant staple food in much of Latin America and Africa. Although some varieties may contain appreciable quantities of protein, its quality is poor because of low lysine and tryptophan content. In 1964, it was discovered that corn bearing a gene known as opaque-2 contained increased concentrations of lysine and tryptophan and had significantly improved nutritional quality. However, opaque-2 corn proved to have low yields, increased susceptibility to diseases and pests, and inferior functional characteristics. At the International Maize and Wheat Improvement Centre in Mexico, work with the opaque-2 gene continued using both traditional breeding and molecular methods.

After at least 12 years' work, CIMMYT researchers succeeded in developing hardy corn varieties that contained twice the lysine and tryptophan content as traditional varieties, but were disease-resistant and high-yielding. Scientists Surinder K. Vasal and Evangelina Villegas of CIMMYT were awarded the World Food Prize in 2000 for their work developing 'Quality Protein Maize'. Quality Protein Maize varieties have been adapted to and released in over 40 countries in Latin America, Africa, and Asia. Recent researchers at CIMMYT reported the development of transgenic corn with multiple copies of the gene from amaranth that encodes for the seed storage protein amarantin. Total protein in the transgenic corn was increased by 32% and some essential amino acids were elevated 8% to 44%. In 2004, a team of researchers at the University of California, Riverside, reported that transgenic corn with increased production of the plant regulating hormone, cytokinin, had nearly twice the content of protein and oil as conventional corn.

This development resulted from an unusual change in the way the plant developed. Normally, corn ears develop flowers in pairs, one of which usually dies. Under the influence of the additional cytokinin, both flowers developed but yielded only a single kernel. These kernels contained more protein and oil than conventional corn. Pursuing a different strategy to improve protein quality, researchers at Monsanto, St. Louis, MO, used genetic engineering techniques to reduce the amount of zein storage proteins. These storage proteins constitute over half the protein in corn and are deficient in lysine and tryptophan. Increased production of other proteins in the corn led to higher levels of lysine, tryptophan, and methionine. The agronomic and nutritional properties of these lines are currently being evaluated. Researchers at the Max Planck Institute, Germany, have focused on methionine, another limiting amino acid. They elucidated several key steps in methionine metabolism in plants.

This work, currently in the preliminary stage, could pave the way for using genetic engineering techniques to improve the methionine content of plants. *Potato*: Potato is a dietary staple throughout parts of Asia, Africa, and South America. Typically, potatoes contain about 2% protein and 0.1% fat. It was reported in 2000 that, as in cassava, transfer of the gene for seed albumin protein from *Amaranthus hypochondriacus* to potato resulted in a "striking" increase in protein content of the transgenic potatoes. In 2004, researchers at the National Centre for Plant Genome Research, India, reported the development of a nutritionally improved potato line with 25% higher yields of tubers and 35%–45% greater protein content. Dubbed the "protato," the protein-rich potato had significant increases in lysine and methionine, which enhance the quality of the additional protein. In February 2004, this potato was reported "approaching release" to farming communities. It should be noted that while potatoes are known for their high starch content, it has been possible to genetically engineer potatoes that contain fat. In July 2004, Klaus and colleagues at the Max Planck Institute of Molecular Plant Physiology demonstrated increased fatty acid synthesis in potatoes. *Rice*: Almost half the world's population eats rice, at least once a day. Rice is the staple food among the world's poor, especially in Asia and parts of Africa and South America. It is the primary source of energy and nutrition for millions.

Thus, improving the nutritional quality of rice could potentially improve the nutritional status of nearly half the world's population, particularly its children. Commodity rice contains about 7% protein, but some varieties, notably black rice, contain as much as 8.5%. The most limiting amino acid in rice is lysine. Efforts to increase the nutritional value of rice target protein content and quality along with key nutrients often deficient in rice-eating populations, such as vitamin A and iron.

The International Rice Research Institute, Philippines, is a primary centre for rice research and development of improved varieties. In 1999, Dr. Momma and colleagues at Kyoto University, Japan, reported a genetically engineered rice having about 20% greater protein content compared with control rice. Transgenic plants containing a soybean gene for the protein glycinin contained 8.0% protein and an improved essential amino acid profile compared with 6.5% protein in the control rice. Dr. William Folk and his team genetically modified rice to increase its content of the amino acid lysine. They did so by modifying the process of protein synthesis, rather than by gene transfer or the expression of new proteins. They achieved an overall 6% increase in lysine content in the grain. Although lysine content remained below optimum levels, the scientists suggested that additional transformations and modifications could further boost lysine levels. Perhaps the most famous genetic transformations in rice are those in "Golden Rice" involving the vitamin A precursor, beta-carotene, and iron.

The lead scientist in the golden rice project, Dr. Ingo Potrykus, now retired from the Swiss Federal Institute of Technology, was also involved in applying biotechnology for the improvement of rice protein. Although details are sparse, Potrykus described the work of Dr. Jesse Jaynes, who synthesized a synthetic gene coding for an ideal high-quality storage protein with a balanced mixture of amino acids. The gene, named Asp-1, was transferred to rice with the appropriate genetic instructions for its production in the endosperm or starchy part of the rice grain.

The transgenic rice plants accumulated the Asp-1 protein in their endosperm in a range of concentrations and provided essential amino acids but data are not yet available on the concentrations achieved or their nutritional relevance. Precedent for the expression of a synthetic gene in rice grown in cell culture suggests that Jaynes' approach is viable.

BIOTECHNOLOGY FOR THE 21ST CENTURY

Experts in United States anticipate the world's population in 2050 to be approximately 8.7 billion persons. The world's population is growing, but its surface area is not. Compounding the effects of population growth is the fact that most of the earth's ideal farming land is already being utilized. To avoid damaging environmentally sensitive areas, such as rain forests, we need to increase crop yields for land currently in use. By increasing crop yields, through the use of biotechnology the constant need to clear more land for growing food is reduced.

Countries in Asia, Africa, and elsewhere are grappling with how to continue feeding a growing population. They are also trying to benefit more from their existing resources. Biotechnology holds the key to increasing the yield of staple crops by allowing farmers to reap bigger harvests from currently cultivated land, while preserving the land's ability to support continued farming.

Malnutrition in underdeveloped countries is also being combated with biotechnology. The Rockefeller Foundation is sponsoring research on "golden rice", a crop designed to improve nutrition in the developing world. Rice breeders are using biotechnology to build Vitamin A into the rice. Vitamin A deficiency is a common problem in poor countries. A second phase of the project will increase the iron content in rice to combat anemia, which is widespread problem among women and children in underdeveloped countries. Golden rice, expected to be for sale in Asia in less than five years, will offer dramatic improvements in nutrition and health for millions of people, with little additional costs to consumers.

Similar initiatives using genetic manipulation are aimed at making crops more productive by reducing their dependence on pesticides, fertilizers and irrigation, or by increasing their resistance to plant diseases.

Increased crop yield, greater flexibility in growing environments, less use of chemical pesticides and improved nutritional content make agricultural biotechnology, quite literally, the future of the world's food supply.

SCOPE

As biotechnology has become widely used, questions and concerns have also been raised. The most vocal opposition has come from European countries. One of the main areas of concern is the safety of genetically engineered food. In assessing the benefits and risks involved in the use of modern biotechnology, there are a series of issues to be addressed so that informed decisions can be made. In making value judgments about risks and benefits in the use of biotechnology, it is important to distinguish between technology-inherent risks and technology-transcending risks. The former includes assessing any risks associated with food safety and the behaviour of a biotechnology-based product in the environment. The latter involve the political and social context in which the technology is used, including how these uses may benefit or harm the interests of different groups in society.

The health effects of foods grown from genetically engineered crop depend on the composition of the food itself. Any new product may have either beneficial or occasional harmful effects on human health. For example, a biotech-derived food with a higher content of digestible iron is likely to have a positive effect if consumed by iron-deficient individuals. Alternatively, the transfer of genes from one species to another may also transfer the risk for exposure to allergens. These risks are systematically evaluated by FDA and identified prior to commercialization. Individuals allergic to certain nuts, for example, need to know if genes conveying this trait are transferred to other foods such as soybeans. Labeling would be required if such crops were available to consumers.

Among the potential ecological risks identified are increased weediness, due to cross- pollination from genetically modified crops spreads to other plants in nearby fields. This may allow the spread of traits such as herbicide-resistance to non-target plants that could potentially develop into weeds. This ecological risk is assessed when deciding if a plant with a given trait should be released into a particular environment, and if so, under what conditions.

Other potential ecological risks stem from the use of genetically modified corn and cotton with insecticidal genes from Bacillus thuringiensis (Bt genes). This may lead to the development of resistance to Bt in insect populations exposed to the biotech-derived crop. There also may be risks to non-target species, such as birds and butterflies, from the plants with Bt genes. The monitoring of these effects of new crops in the environment and implementation of effective risk management approaches is an essential component of further research. It is also important to keep all risks in perspective by comparing the products of biotechnology and conventional agriculture.

The reduction of biodiversity would represent a technology-transcending risk. Reduced biological diversity due to destruction of tropical forests, conversion of land to agriculture, overfishing, and the other practices to feed a growing world population is a significant loss far more than any potential loss of biodiversity due to biotech-derived crop varieties. Improved governance and international support are necessary to limit loss of biodiversity. What we know from our understanding of science and more than a decade of experience with biotech-derived plants is the following: There is no evidence that genetic transfers between unrelated organisms pose human health concerns that are different from those encountered with any new plant or animal variety. The risks associated with biotechnology are the same as those associated with plants and microbes developed by conventional methods.

CONSUMER AND FOOD INDUSTRY PERSPECTIVES

Survey research over the past decade shows that biotechnology is not likely to become an important issue for most American consumers. Consumers find biotechnology acceptable when they believe it offers benefits and it is safe. Surveys have consistently found that a majority of American consumers are willing to buy insect-protected food crops developed through biotechnology that use fewer chemical pesticides, as well as more nutritious foods. American consumers also appreciate the role that biotechnology can play in feeding the world. Research shows that European consumers are much less supportive of all biotechnology applications.

Surveys since 1992 show that relatively few U.S. consumers have heard or read much about biotechnology. News about the cloned sheep pushed awareness to 50 per cent in March 1997. Surveys in the first three months of 2000 show that awareness has fallen back to just over one-third in the United States. Such trends reflect the fact that most people get their information about biotechnology from the media. Unfortunately, many consumers also do not understand some fundamental principles of biology. European consumer awareness is somewhat higher, but knowledge is still low.

Media coverage in the United States has generally been balanced (which helps account for our relatively high levels of acceptance). This is in sharp contrast to the European media, which have played upon fear of the unknown. The European media have also tended to accept opponents' claims without question. Another issue is that many people no longer have a connection to agriculture. In fact, research has shown that many consumers are unaware that all foods are derived from plants or animals that already have been genetically modified through traditional (but imprecise) breeding methods.

American consumers look to health professionals and scientific experts for credible information, but place relatively little trust in the activists who oppose biotechnology. Research shows that acceptance increases significantly when American consumers learn that organizations such as the National Academy of Sciences and the U.S. Food and Drug Administration have determined that biotech-derived foods are safe. In contrast, European consumers express the most trust in those groups that oppose biotechnology. They have much less confidence in government, industry, or even scientists. American culture is more supportive and rewarding of new technology. Europeans tend to view food differently from U.S. consumers. In fact, some Europeans reject all American food products. Europeans also want to protect their small farms to maintain open space and rural employment. Such forces underlie much of the European anxiety about agricultural biotech - especially since it is seen as an "American invention." Most of the industry leaders interviewed are quite enthusiastic about the benefits of biotechnology — especially in terms of increased food availability, enhanced nutrition, and environmental protection. Most feel that biotechnology has already provided benefits to consumers. Almost all recognize that foods developed through biotechnology have already been part of consumers' everyday diet. They clearly do not agree with most of the opponents' claims and tend to have almost no trust in such groups.

Their main concerns involve lack of consumer acceptance — not the safety of the foods. They express high levels of confidence in the science and the regulatory process. In fact, almost none feel that biotechnology should not be used because of uncertain, potential risks. Most food industry leaders do not feel it is necessary to have special labels on biotech-derived foods. They express concerns that such labels would be perceived as a warning by consumers. They also worried that the need to segregate commodities would pose financial and logistical burdens on everyone in the system - including consumers. Food industry leaders recognize a major need to educate the public about biotechnology. They look to third parties, such as university and government scientists to provide such leadership.

Research shows that consumers will accept biotech foods if they see a benefit to themselves or society and if the price is right. Their responses to foods developed through biotechnology are basically the same as for any other food - taste, nutrition, price, safety and convenience are the major factors that influence our decisions about which foods to eat. How seeds and food ingredients are developed will only be relevant for a relatively small group of concerned, consumers.

The potential for public concerns has led several food companies to change their products to avoid biotech-derived ingredients. For example, Gerber Foods received threats from Greenpeace because they had determined the company was using biotech-derived food ingredients (mainly soy). The company firmly believes that the biotech foods are safe to consume. Gerber agreed to drop some of its existing corn and soybean suppliers in favour of ones that can produce crops that are not genetically altered. It became an issue that is suddenly confronting other food companies. A private manufacturer in California, called Healthy Times Natural Food has switched from Canola oil (which sometimes is genetically modified) to safflower oil after facing questions from Greenpeace. The controversy is due in part to the fact that the organic industry is using public concern as a tool for marketing their products as free of biotech ingredients.

NATIONAL AND INTERNATIONAL BIOTECHNOLOGY POLICY

National governments and international policy making bodies rely on food scientists and others to develop innovations that will create marketable food products and increase food supplies. Governments also rely on scientific research because they are responsible for setting health and safety standards regarding new developments. International organizations can suggest policy approaches and help develop international treaties that are ratified by national governments.

Economic success in the competitive international market demands that food production become more efficient and profitable. National governments and international organizations support food biotechnology as a means to avoid global food shortages. Many policy making bodies are also trying to balance support of the food biotechnology industry with public calls for their regulation. Such regulations are necessary to protect public health and safety, to promote international trade, conserve natural resources, and account for ethical issues.

The majority of processed foods on the market contain soy or corn ingredients that come from GM plants. To date none have posed a food safety risk. The chief safety concerns are the potential to alter nutrient content or introduce allergens. Federal agencies involved in biotechnology regulation include the U.S. Department of Agriculture (USDA) which evaluates agricultural production processes for all foods; the Food and Drug Administration (FDA), which evaluates whole non-animal foods (seafood), food ingredients, and food additives; and the Environmental Protection Agency (EPA), which evaluates plants with insecticidal properties.

Developers of GM plants and biotech-derived foods are required to consult with FDA prior to the commercialization of the product. This consultation procedure entails a science-based safety assessment of the product that focuses on the protection of the consumer, developer, and the environment. Thus developers, have a strong incentive to cooperate fully with FDA and the other agencies prior to marketing their products.

FOOD LABELING

Labeling food derived from GM plants and animals is an important, but complex issue. Some consumers and consumer groups believe they have a right to know whether biotechnology was used to produce food. Others believe labeling is not necessary if foods are essentially equivalent in composition. Food labels are regulated by FDA and, in some cases, by USDA. Regulatory agencies are concerned with ensuring that food labels are both true and not misleading. In accordance with their statutory mandate, the FDA has determined that a food product should be labeled as a product of biotechnology only if it has been changed in some significant way. The U.S. Court of Appeals recently reaffirmed that the FDA's labeling policy is correct.

Based on the input from three public hearings, the FDA has recognized that the most effective way to allow for informed choice will be a system of voluntary labeling for foods not produced through biotechnology. If the demand for "GM-free" food is real, a market will become viable. In this case, meaningful choice can be provided to concerned consumers without imposing costs on or denying benefits to the majority of consumers, who support biotechnology. This may be the only viable and cost-effective option for the food industry (given the costs and difficulties associated with segregation and identity preservation.)

National surveys conducted with American consumers between 1997 and 2000 found about three-quarters of consumers supported this FDA labeling policy. There is evidence from focus groups that US consumers are already overwhelmed by the level of detail on food labels and mainly look for relevant information about nutrition. Recent focus groups have demonstrated that the wording on labels has a significant effect on consumer understanding and acceptance of biotechnology. Any label messages need to be simple and clear. However, labeling will be counterproductive without education.

It is important to understand when FDA will and won't require a product label. The same labeling laws that apply to all other foods and food ingredients apply to products of food biotechnology. Requirements for all food labels mandate proper identification of products and notice of health or safety concerns. Labeling would be required in the following instances - but not simply because the products were made using biotechnology.

- Potential food allergy is an example of a health or safety risk that would mandate a product label (*e.g.*, if a gene has been transferred from a known allergen, such as a peanut).
- When substantial changes have been made to the expected nutritional composition of a food (*e.g.*, vitamins have been added to a particular type of produce).
- If the common identity of a food product has been changed (*e.g.*, if the composition of a fruit or vegetable becomes significantly different it would have to have a different name).

Labeling of biotech-derived foods will be required in some countries (such as Japan and the European Union) for cultural and religious reasons. These countries also believe labeling should be required simply because the consumers want to know more about the content of the food or how it was produced. Some have suggested that such labeling initiatives have been driven by economic and political interests rather than by scientific need. Consumers in these countries will ultimately end up paying even higher prices for their food as a result of such strict labeling policies.

The applications of biotechnology are so broad, and the advantages so compelling, that virtually every industry is using this technology. Developments are underway in areas as diverse as pharmaceuticals, diagnostics, textiles, aquaculture, forestry, chemicals, household products, environmental cleanup, food processing and forensics to name a few. Biotechnology is enabling these industries to make new or better products, often with greater speed, efficiency and flexibility. Biotechnology holds significant promise to the future but certain amount of risk is associated with any area. Biotechnology must continue to be carefully regulated so that the maximum benefits are received with the least risk.

Biotechnology is at a crossroads in terms of public acceptance. Many Americans have not yet formed a solid opinion on this complex issue. International developments over the next few years will certainly have a major influence on the long-term viability of biotechnology. The future of the world food supply depends upon how well scientists, government, and the food industry are able to communicate with consumers about the benefits and safety of the technology.

4

Evaluation of Genetically Modified Food Crops

Genetic modification, otherwise referred to as recombinant DNA (rDNA) technology or gene-splicing, has proven to be a more precise, predictable and better understood method for the manipulation of genetic material than previously attained through conventional plant breeding. To date, agricultural applications of the technology have involved the insertion of genes for desirable agronomic traits (e.g. herbicide tolerance, insect resistance) into a variety of crop plants, and from a variety of biological sources.

Examples include soybeans modified with gene sequences from a Streptomyces species encoding enzymes that confer herbicide tolerance, and corn plants modified to express the insecticidal protein of an indigenous soil microorganism, Bacillus thuringiensis (Bt). A growing body of evidence suggests that the technology may be used to make enhancements to not only the agronomic properties, but the food, nutritional, industrial and medicinal attributes of genetically modified (GM) crops.

Regulatory supervision of rDNA technology and its products has been in place for a longer period of time in the United States than in most other parts of the world. The methods and approaches established to evaluate the safety of products developed using rDNA technology continue to evolve in response to the increasing availability of new scientific information.

As our understanding of the potential applications of the technology is broadened, the safety of products developed using rDNA technology and the potential effects of introduced gene sequences on human health or the environment will be more closely scrutinized. In fact, much of the knowledge acquired during the commercialization of the products of rDNA technology in agriculture is now finding application in evaluating the safety of products developed through more conventional means. The objective of this chapter is to provide the reader with an overview of the significant events leading up to the present, science-based, regulatory framework that exists for the safety evaluation of GM food crops within the United States. An attempt has been made to discuss concerns over the sufficiency of existing regulations, as well as to highlight recent initiatives taken by federal regulatory agencies to address them. Through better communication of how the regulatory process functions within the United States, it is anticipated that current and future applications of rDNA technology in agriculture will be met with a greater level of understanding and acceptance.

HISTORICAL PERSPECTIVE

rDNA technology was first developed in the 1970s. The initial response of the scientific community, including members of the National Academy of Science (NAS), to the prospects of rDNA technology, was to postpone any further research involving the technology until the potential risks to human health and the environment could be evaluated.

Researchers attending the International Conference on Recombinant DNA Molecules in 1975, otherwise known as the Asilomar Conference, tried to establish a scientific consensus on how best to self-regulate emerging applications of the technology. The conditions and restrictions that were proposed at this conference have formed the basis by which federal guidelines and policies for rDNA technology research were drafted within the United States.

National Institutes of Health (NIH)

The National Institutes of Health (NIH) was the first federal regulatory agency to publish their interests in evaluating the safety of rDNA technology in 1976, in the form of guidelines for the conduct of research. Because of the uncertainties that existed at the time, all research into the potential applications of rDNA technology was limited to the confines of federally funded labouratories under NIH control. After continued research, and a more careful assessment and monitoring of the risks, a set of less restrictive guidelines was published in 1978.

However, the environmental release of organisms developed using rDNA technology outside the confines of controlled labouratory conditions was prohibited unless otherwise approved by the NIH director. In the early 1980s, the NIH established an rDNA Advisory Committee (RAC) to review all data and experience gained with applications of the technology under its control. Based on recommendations of the RAC, a more relaxed set of research guidelines was published by the NIH in 1983.

The NIH approved the first environmental release of an organism developed using rDNA technology (ice-minus strain of Pseudomonas) in 1983. In response, they were criticized for failing to prepare a statement or assessment of the environmental impact of their regulatory decision as required under the National Environmental Policy Act (NEPA). Once the legal controversy had subsided, all responsibility that the NIH had for regulating the environmental introduction of GM organisms was relinquished.

Nevertheless, NIH guidelines continue to be referenced in assessing the safety of rDNA research performed within industry, federal and other state labouratories. However, it was unclear which federal regulatory agencies would be responsible for ensuring the safety of the products developed using rDNA technology.

OFFICE OF SCIENCE AND TECHNOLOGY POLICY (OSTP)

In response to a need for clarification, the Office of Science and Technology Policy (OSTP) began work on the development of a policy to establish a federal regulatory framework for evaluating the safety of products developed using rDNA technology. Following an opportunity for public comment, OSTP published a final version of the 'Coordinated Framework for Regulation of Biotechnology' (Coordinated Framework) in 1986.

The policy provided the basis by which federal regulatory agencies got involved in evaluating the safety of products at later stages of commercial development at that time. According to the Coordinated Framework, the products of rDNA technology should be regulated on the basis of the unique characteristics and features that they exhibit, not their method of production. The products of rDNA technology were considered to pose risks to human health and the environment similar to those posed by conventional products already regulated within the United States.

As a result, no new federal regulatory agencies or regulations were required. The Coordinated Framework did not, however, rule out the possibility of the development of new guidelines, procedures, criteria or even regulations to supplement or alter the scope of existing statutes for the products of rDNA technology. The Coordinated Framework identified three federal regulatory agencies within the United States: the US Food and Drug Administration (US FDA), the US Department of Agriculture (USDA) and the US Environmental Protection Agency (US EPA), as having primary responsibilities for evaluating the products of rDNA technology under development at that time. In 1992, the OSTP released another document entitled, 'Exercise of Federal Oversight within the Scope of Statutory Authority: Planned Introductions of Biotechnology Products Into the Environment', outlining the proper basis by which federal regulatory agencies were expected to exercise their regulatory authority. As with conventional products, dependent upon the intended use and function, more than one federal regulatory agency may share an interest in evaluating the safety of a product developed using rDNA technology.

If more than one federal regulatory agency has an interest, lead agencies are identified as being responsible for coordinating activities to limit any potential duplication of efforts. Although federal regulatory agencies worked independent of one another, it was realised that close working relationships would need to be established in order to evaluate effectively the safety of products developed using rDNA technology. Recently, the OSTP teamed up with the White House Council on Environmental Quality (CEQ) to perform a six-month inter-agency evaluation of the federal regulatory agency responsibilities in evaluating the environmental safety of products developed using rDNA technology. A case-study approach for a variety of different classes of products developed using rDNA technology was used to evaluate the level of federal regulatory agency involvement, to identify strengths, weaknesses and areas of potential improvement. The review concluded that none of the previously approved products of rDNA technology has had any significant negative impact on the environment.

Although all the case studies were published, OSTP/CEQ failed to reach a consensus on issues relating to the relevant strengths and weaknesses of the existing regulatory structure within the time allotted for the completion of its review. A review of the case studies published provides a comprehensive interpretation of the responsibilities of each federal regulatory agency in ensuring the safety of the products developed using rDNA technology.

NATIONAL ACADEMY OF SCIENCES (NAS)

The National Academy of Sciences (NAS), and its operating arm, the National Research Council (NRC), have served as a primary source of scientific, technological, human health and environmental policy advice during the development of regulatory approaches for the safety evaluation of products developed using rDNA technology within the United States. In 1987, the NRC published a report concerning the potential human health and environmental hazards associated with the commercial introduction of GM organisms, entitled Introduction of Recombinant DNA-Engineered Organisms into the Environment: Key Issues. The risks associated with the introduction of GM organisms were considered to be essentially the same in kind as those associated with unmodified organisms.

In other words, rDNA technology did not appear to introduce any unique risks as compared to the products that had been developed using more conventional methods of genetic modification. In reaching these conclusions, the NRC performed an evaluation of the similarities and differences in the properties exhibited by products developed using a variety of different techniques. To this day, the conclusions of this report continue to be referenced by the regulators and developers of GM crop varieties worldwide. A subsequent NRC report, entitled Field Testing Genetically Modified Organisms: Framework for Decisions, also reached similar conclusions, but provided additional guidance as to how regulatory decisions concerning the introduction of GM organisms should be made. The NRC recommended that regulatory decisions concerning the introduction of GM organisms should be made on a case-by-case basis. Consistent with the Coordinated Framework, the NRC did not consider the nature of the process used for the genetic modification of an organism to be a useful criterion for determining whether a product requires less or more regulatory oversight. As a result, no valid reason existed to regulate organisms genetically modified via modern techniques (e.g. rDNA technology) any differently from organisms genetically modified via more conventional means.

Similar conclusions have been published in the reports of international standards-setting organizations. In retrospect, within both reports, the NRC acknowledged that modern and conventional methods of genetic modification are not without risks to human health or the environment, and as a result, neither could be considered inherently more risky. As a result, regulatory decisions concerning the safety of the products of rDNA technology need only to take into consideration the specific characteristics exhibited by a particular GM organism and the environment in which it is to be introduced, and not the method by which it has been produced.

The NRC has further articulated their conclusions into what is now commonly referred to as the 'Concept of Familiarity'. Although familiarity with the characteristics of a particular organism or the environment to which it will be introduced would not necessarily mean it was safe, it can be expected to provide a sufficient amount of information to allow for a judgement to be made of the risks.

For example, familiarity with a new GM plant variety could be established based on comparisons between characteristics of the parent line or other crop species exhibiting similar traits, as well as through the results of actual field tests involving the GM plant. These principles were further elabourated upon by the Organisation for Economic Co-operation and Development (OECD), and as a result, have been referenced in the development of regulatory policies for evaluating the safety of GM crops on a global basis.

More recently, a committee established by the NRC published a report entitled Genetically Modified Pest-Protected Plants: Science and Regulation, based on a review of all scientific and regulatory data collected during the regulatory approval process for GM crops within the United States. The primary objective was to assess independently the effectiveness of existing and proposed regulations for the safety evaluation of GM crops expressing plant pesticides (i.e. plant-incorporated protectants).

No new evidence was identified to suggest plants expressing plant pesticides posed any greater risk to human health or the environment as a result of their genetic modification. In fact, the NRC concluded, 'with careful planning and appropriate regulatory oversight, commercial cultivation of GM plants is not expected to pose higher risks and may pose less risk than other commonly used chemical and biological pest-management techniques'.

However, the NRC report included requests for federal regulatory agencies to further strengthen the current regulatory approval process through better coordination and communication between agencies, on-going investment in the research and monitoring of potential human health (e.g. allergenicity) and environmental impacts (e.g. insect resistance), and by providing greater access to information evaluated in support of regulatory decisions.

DEFINITION AND SCOPE OF BIOENGINEERED FOODS

The recently proposed rule of the US FDA concerns 'bioengineered foods' which have been defined as 'foods derived from plant varieties that are developed using in vitro manipulations of DNA (generally referred to as rDNA technology)'. As a result, the proposed rule has a much narrower focus than the 1992 US FDA Statement of Policy.

The US FDA has explained the need for a change in emphasis based on their expectations that many of the new plant varieties exhibit a greater potential to 'contain substances that are significantly different from, or that are present in food at a significantly different level than before'. As a result, the substances present in foods and food components derived from new plants developed using rDNA technology are less likely to be considered GRAS, and as a result, will require pre-market approval from the US FDA.

SAFETY AND NUTRITIONAL EVALUATION

The primary objective of the safety and nutritional evaluation is to demonstrate that the food derived from a new plant variety is as safe or nutritious as foods already consumed as a part of the diet. For new plant varieties, including those developed using rDNA technology, a science-based approach is used to focus the evaluation on the demonstrated characteristics of the food or food component.

The evaluation of a GM food or food component typically involves reviewing information or data on any newly introduced substances, the known levels of toxicants, as well as the nutritional composition of the plant following modification. Substances that raise safety concerns (e.g. toxicants, allergens) would be subject to more extensive evaluation, since both intended and unintended changes may affect the levels of toxicants and nutrients in a food following the modification.

Guidance for performing a safety and nutritional evaluation was provided in the 1992 US FDA Statement of Policy, in a series of flow charts and text that cover:

- The crop that has been modified;
- Source(s) of the introduced genetic material;
- New substances intentionally added to the food as a result of the genetic modification (e.g. proteins, but also fatty acids, and carbohydrates).

Documentation required to support the evaluation typically includes: the purpose or intended technical effect of the modification on the plant, together with a description of the various applications or uses, a molecular characterization of the modification including the identities, sources and functions of introduced genetic material; information on the expressed protein products encoded by introduced genes; information relating to the known or suspected allergenicity and toxicity of any expressed gene products; for foods known to cause allergy, information on whether the endogenous allergens have been altered by the genetic modification; information on the compositional and nutritional characteristics of the foods, including anti-nutrients; and in some instances, comparative results of feeding studies involving the foods derived from plants modified using rDNA technology and the non-modified counterpart. In performing its evaluation, the US FDA is particularly interested in the identification of inherent toxicants, known or potential allergens, assessing the concentration and bioavailability of essential nutrients, the safety and nutritional value of any newly introduced proteins, and the identity, composition, and nutritional value of modified carbohydrates, fats and oils. If additional questions of safety remain following this evaluation, further toxicological studies may need to be performed.

It is recognized that absolute assurance of the safety of any food does not exist. As a result, the goal of the safety evaluation is to establish a reasonable certainty of no harm under anticipated conditions of consumption. With this in mind, experience with the existing food supply has provided the basis for evaluating the safety of new food or food components. Both the Food Advisory Committee and Committee for Veterinary Medicine have been extensively involved in the development of approaches for the safety and nutritional evaluation of foods and food components derived from new plant varieties, including those developed using rDNA technology.

PRODUCT CHARACTERIZATION

Product characterization takes into consideration information relating to the modified food crop, the introduced genetic material and its expression product, and acceptable levels of inherent plant toxicants and nutrients. All characteristics of the gene insert must be known, including the source(s), size, number of insertion sites, promoter regions, and marker sequences.

It must be established that the transferred genetic material does not come from a pathogenic source, a known source of allergens, or a known toxicant-producing source. The introduced genetic material should be well characterized to ensure that the introduced gene sequences do not encode harmful substances and are stably inserted within the plant genome to minimize any potential opportunity for undesired genetic rearrangement. Analytical data are required to evaluate the nutritional composition, the levels of any known toxicants, anti-nutritional and allergenic substances, and the safety-in-use of antibiotic resistance marker genes. Any new substances introduced into crops through rDNA technology (e.g. proteins, fatty acids, carbohydrates) will be subject to pre-market review as food additives by the US FDA, unless substantially similar to substances already safely consumed as a part of foods, or that are considered GRAS. To date, substances that have been added to foods through rDNA technology have been previously consumed or have been determined to be substantially similar to substances already consumed as a part of the diet. As such, introduced substances have been considered exempt from the requirement for pre-market approval as food additives with the US FDA. A more rigorous safety evaluation of a GM crop is warranted if the introduced gene sequence(s) has not been fully characterized, the nutritional composition has been significantly altered, antibiotic resistance marker genes have been used during its development, or if an allergenic protein or toxicant has been detected at levels higher than what is typically observed in edible varieties of the same crop species. In any event, determinations as to the safety of substances that have been introduced into new plant varieties through rDNA technology are made on a case-by-case basis.

Although an evaluation of the introduced gene sequences and expression product(s) provides assurance as to their safety, further studies may be required to predict whether unexpected effects may result following their interaction with other genes within the plant. In addition, the product characterization of a GM plant involves assessing sequence homology to known toxicants and allergens, thermal and digestive stability, and if required, the results of both in vitro and in vivo assays to demonstrate lack of toxicity.

COMPOSITIONAL ANALYSIS

The results of field trials performed over several years serve to characterize the phenotypic and agronomic characteristics exhibited by the plant (e.g. height, colour, leaf orientation, susceptibility to disease, root strength, vigour, fruit or grain size, yield, etc.), as well as to provide the materials required for the compositional analysis. Any anomalies in the phenotypic or agronomic characteristics exhibited by a plant may result in a requirement for additional information. Protein, fat, fibre, starch, amino acid, fatty acid, ash and sugar levels are determined, as well as the levels of anti-nutrients, natural toxicants or known allergens.

Studies of the nutritional composition are performed to determine whether the levels of any key nutrients, vitamins or minerals have been altered as a result of the genetic modification. Based on the results of these studies, a determination is made as to whether the phenotypic and agronomic characteristics of a GM crop or the concentrations of inherent constituents fall within ranges typical of its conventional counterpart.

If inserting a new gene causes no change in any of the assessed parameters, the US FDA can conclude with reasonable assurance that the GM crop is as safe as the conventional crop. If the levels of essential nutrients or inherent toxicants are found to be significantly different in the GM crop, the US FDA may recommend additional action prior to commercialization, such as obtaining food additive status, or the use of specific labels to alert consumers of an altered nutritional content, etc.

ALLERGENICITY

In consultation with scientific experts in the areas of food safety, food allergy, immunology, biotechnology and diagnostics, the US FDA published guidelines for assessing the allergenicity of GM foods or food components in 1994. The approach to assessment is multi-faceted, incorporating data regarding the origin of the genetic material, and the biochemical, immunological and physicochemical properties of the expressed protein.

The overall assessment is reliant upon the fact that all known food allergens are proteins and, notwithstanding the number of shared properties between allergenic and non-allergenic food proteins, food allergens tend to exhibit a number of similar characteristics. In general, food allergens share a number of common properties: they have a molecular weight of over 10 000 Da; they represent more than 1% of the total protein content of the food; they demonstrate resistance to heat, acid treatment, proteolysis and digestion; and they are recognized by IgE.

For gene sequences derived from known allergenic sources (e.g. peanuts), the developers of GM plants are expected to demonstrate that allergenic proteins have not been introduced into the food. For assessment purposes, it is assumed that any genetic material derived from a known allergenic source will encode for an allergen. To demonstrate otherwise, the amino acid sequence of an expressed protein must be compared with that of known allergens using protein sequence databases.

Furthermore, in vitro and/or in vivo immunologic analyses using the sera of allergic patients sensitive to the source of the genetic material may need to be performed to determine whether or not a potentially allergenic protein is being expressed in the GM food. Some GM foods may be modified to express genes from a source that is not known to be allergenic when consumed.

Under these circumstances, the US FDA follows a similar decision tree-based approach to determining the allergenic potential of the expression product. In assessing these proteins, should any amino acid sequence exhibit homology with a known allergen, the expressed protein would then be evaluated in immunologic tests using the sera of patients known to be allergic to the identified homologous protein. Regardless of the origin of the genetic material, physicochemical studies are performed in vitro to provide information concerning the expected stability of the expressed protein.

All known food allergens tend to be resistant to digestive degradation, as demonstrated in simulated gastric fluid models, or to decomposition under conditions of food processing. In making a determination regarding a GM food, it is the totality of the biochemical, immunological and physicochemical properties of the introduced protein that provides guidance as to the allergenic potential of such a protein being expressed in food.

The level of protein expressed, produced and consumed as a part of the diet is also a primary indicator of the allergenic potential, since nearly all food allergens are known to be major proteins in their respective foods. If the results of any of these studies suggest an allergenic potential, the US FDA may recommend further scientific evaluation, require special labelling to alert sensitive consumers, or alternatively caution the developer about proceeding with the development of a particular GM food.

Most recently, a joint FAO/WHO Expert Consultation on Allergenicity of Foods Derived from Biotechnology recommended a revised decision tree. This decision-tree strategy was modified from the previous version FAO/WHO to include a revised definition of sequence homology for gene product comparison; a greater emphasis on serum testing, even with gene products without homology to known allergens and not derived from an allergenic source; and animal models to assess potential allergenicity, despite acknowledgement by the Expert Consultation that these models are currently under development and at present, not predictive of food allergies in humans.

More recently, the ad-hoc Open-Ended Working Group on Allergenicity established by the ad-hoc Intergovernmental Codex Task Force on Foods Derived from Biotechnology, considered the FAO/WHO strategy in drafting an approach to assessing the potential allergenicity of foods derived from rDNA plants.

The Codex Working Group recognized the absence of a definitive predicitve test for allergenicity in humans to a newly expressed protein and recommended an integrated, stepwise assessment strategy.The strategy recommended by the Working Group, and accepted by the Codex Task Force for inclusion in the draft forwarded for final adoption by the Codex Alimenarius Commission, is consistent with that of FAO/WHO. However, the Working Group suggested that some of the modifications included in FAO/WHO could contribute to the overall weight of evidence of any conclusion of potential allergenicity (e.g., allergen specific serum depositories, animal models), pending development and validation.

ANTIBIOTIC RESISTANCE

The use of antibiotic-resistance genes as selectable markers has been common practice in the development of new plant varieties using rDNA technology. Concerns relate to the potential transfer of antibiotic-resistance genes from GM plants to pathogens in the environment or to the gut of humans consuming foods or food components. Issues relating to the use of antibiotic resistance genes were identified in the 1992 US FDA Statement of Policy Statement as well as discussed in additional guidance entitled Guidance for

Industry: Use of Antibiotic Resistance Marker Genes in Transgenic Plants. The guidance provided within these documents was established in consultation with experts in the fields of microbiology, medicine, food safety, bacterial and mycotic diseases, and includes suggestions with respect to the continued safe use of antibiotic-resistance marker genes by the developers of new plant varieties. The use of marker genes that encode resistance to clinically important antibiotics has raised questions as to whether their presence in food could reduce the effectiveness of oral doses of the antibiotic or whether the gene present in the DNA could be transferred to pathogenic microbes, rendering them resistant to treatment with the antibiotic.

The risk of transfer of antibiotic-resistance genes from plants to microorganisms considered to be pathogenic to humans, however, is considered to be minimal if not insignificant. Furthermore, the potential risks are becoming less of a concern as more developers are beginning to research the use of alternative technologies (e.g. non-resistance-based markers) in plant breeding. The conclusions with respect to the safe use of antibiotic-resistance marker genes are consistent with the findings of other national and international food safety organizations.

Consultation and Filing Process

The submission of a Pre-market Biotechnology Notification (PBN) has recently been proposed as a mandatory requirement for the commercialization of bioengineered foods and food ingredients within the United States. Minimal differences exist between the information to be submitted as a part of a PBN and that previously presented in voluntary consultations with the US FDA.

As proposed, developers are still being encouraged to consult with the US FDA as early and as often as necessary in the development of a bioengineered food, such that any potential scientific or regulatory concerns can be identified and addressed prior to the submission of the PBN. Guidelines for performing consultations with the US FDA were released in a publication entitled Guidance On Consultation Procedures: Foods Derived from New Plant Varieties in 1997.

The publication recommended an approach for developers to proceed by submitting a request for consultation, outlined the internal process by which all requests would be handled, and included additional guidance as to the type of safety and nutritional information to be presented during consultation with the US FDA. Once sufficient safety and nutritional information has accumulated to demonstrate that a product is safe and in compliance with the FFDCA, developers typically schedule a consultation to present their scientific findings and conclusions to the US FDA.

Consultations prior to notification not only serve to keep the US FDA informed of advances made in the application of rDNA technology in food production, but also keep the developers of bioengineered foods aware of emerging safety, nutritional or regulatory concerns of the US FDA. Consultations are considered complete when all safety and regulatory concerns between the US FDA and the developer have been resolved.

The US FDA proposes to perform an initial evaluation of a PBN within 15 days of receipt to determine completeness, at which point, if considered complete, the PBN will be filed, and a response can be expected within 120 days.

The US FDA does not issue a product approval per se, but informs the developer by letter that:

- The evaluation period has been extended;
- The notice is not complete and why;
- It has no further questions 'at this time' based on the information that has been presented.

Labelling

The FFDCA defines what information must be disclosed to consumers on a food label, such as the common or usual name, and other limitations concerning the representations or claims that can be made or suggested about a food product. All foods must be labelled truthfully and not be misleading to consumers.

Taking this into consideration, the FFDCA does not stipulate the disclosure of information on the basis of consumer desire to know. Labelling may be considered misleading if it fails to reveal material facts in light of representations that are made with respect to a product. The labelling of foods derived from new plant varieties, including plants developed using rDNA technology, was originally addressed in the 1992 US FDA Statement of Policy, and most recently discussed in Draft Guidance for Industry for the voluntary labelling of bioengineered foods.

To date, the US FDA is not aware of any information that would distinguish foods developed using rDNA technology (e.g. bioengineered foods) as a class from foods developed through other methods of conventional plant breeding, and as such, have not considered the method of development a material fact requiring disclosure on product labels.

Nevertheless, after extensive consultation, including thousands of written comments and a series of public meetings, the US FDA has observed 'a general agreement that providing more information to consumers about bioengineered foods would be useful'.

Requirements for Labelling

Special labelling is required if the composition of the bioengineered food differs significantly from its conventional counterpart. For example, for a food that has been genetically modified to contain a new major sweetener, a new common or usual name or other labelling may be required.

Similarly, if a GM food contains an allergen that consumers would not expect to be present in that food, special labelling may be necessary to alert sensitive consumers. If a protein commonly associated with an allergic reaction (e.g. peanut protein) is transferred to another food through genetic modification, the US FDA would evaluate whether labelling would provide sufficient consumer protection. If labelling would not be considered to provide a sufficient level of protection, the US FDA would take appropriate steps to ensure the GM food would not be marketed.

Therefore, current policy requires a GM food to be labelled when the resulting product poses a safety issue or is substantially different from its conventional counterpart, and as a result, could be considered to pose a misrepresentation to consumers. The 1992 US FDA Statement of Policy and the recent Draft Guidelines do not consider the use of rDNA technology in the development of food products to be a material fact requiring specific disclosure on the label.

Rather it is a method of development, similar to other methods of plant breeding, which have not required disclosure on the label. Bioengineered foods cannot be distinguished compositionally from foods modified through more conventional methods and thus do not require specific disclosure through labelling. The Draft Guidelines reaffirm the US FDA position that bioengineered foods do not require special labelling.

Voluntary Labelling

To provide guiding principles for voluntary labelling, in recognition of the desire of certain manufacturers to label foods as produced either with or without bioengineering, the US FDA published a 'Draft Guidance for Industry: voluntary labelling indicating whether the foods have or have not been developed using bioengineering'.

Emphasizing that the use of rDNA technology 'is not a material fact', the US FDA recognizes that some consumers want disclosure of bioengineered content and that some manufacturers wish to provide it. In response, Draft Guidance was issued with suggestions concerning the use of labelling statements that are not considered misleading.

GENETICAL MODIFIED ORGANISM IN DERIVED FOODS

The concept of substantial equivalence was originally developed through discussions at the Organisation for Economic Co-operation and Development, though, to a large extent, these discussions built on previous work done by the World Health Organization (WHO) and the Food and Agriculture Organization. In 1991, a Joint FAO/WHO Consultation had concluded that the evaluation of a food derived through modern biotechnology should consider both food safety and nutritional value using similar conventional food products as a standard and taking into account the processing of the food and its intended use. In the 1980s, the OECD established a Group of National Experts on Safety in Biotechnology (GNE), which continued to work through to 1993. As an intergovernmental organization, OECD's GNE was attended by delegates nominated by the governments of the OECD member countries, who were, for the most part, representatives of those agencies and ministries with a responsibility for safety in biotechnology. As a result of the work of the GNE, the OECD published a number of important documents relevant to the safety assessment of biotechnology-derived products during this period, which deal with a range of issues, including safety considerations for industrial, agricultural as well as environmental applications of organisms derived by recombinant DNA techniques.

It was recognized at this stage that the safety assessment of an organism derived through recombinant DNA techniques would rely heavily on the knowledge of its parental organism, as well as on an analysis of how the new organism appears to differ from the parent. OECD recommended that the considerable data on environmental and human health effects of living organisms that exists should be used to guide the risk assessment.

By the early 1990s, there had already been large numbers of small-scale field trials of new crop varieties derived through recombinant DNA techniques, and it became clear that new foods derived from these varieties would be marketed during the 1990s. In order to proactively address food safety issues related to these novel foods, in 1990, the GNE established a Working Group on Food Safety and Biotechnology comprising experts, mainly from the ministries and agencies responsible for food safety issues in OECD member countries. The main objective of the Working Group was to elabourate scientific principles and concepts to be used when evaluating the safety of new foods and food components of terrestrial microbial, plant or animal origin.

The Working Group did not consider the safety assessment of food additives, contaminants, processing aids or packaging materials. Nor did it consider environmental safety issues which had been (or were being) addressed by other groups of the GNE. Early in the life of the Working Group, an important recognition was that traditionally, the safety of food for human consumption had been based on the reasonable certainty that no harm will result from intended uses under the anticipated conditions of consumption. Foods prepared and used in traditional ways have usually been considered safe on the basis of long-term experience, even though they may have contained natural toxicants or anti-nutritional substances.

Normally, new varieties of foods or crops have not been subjected to traditional toxicological testing. In fact, where toxicological testing has been applied to whole foods, the results have often been difficult to interpret. Although the OECD's Working Group recognized that modern biotechnology might extend the scope of genetic changes that can be made - and might even broaden the range of the possible sources of foods - it was recognized that this would not inherently lead to foods that are less safe than those developed through conventional techniques.

The Working Group therefore indicated that the evaluation of foods or food components derived through modern biotechnology does not require a fundamental change in established principles, nor does it require a different standard of safety. Realizing the importance of using examples of new foods to identify and demonstrate the applicability of the proposed scientific principles, the Working Group organized a number of meetings and intergovernmental consultations which included case study presentations of novel foods such as enzymes, genetically modified bakers' yeast, mycoprotein and a number of genetically modified varieties of crop species.

Although these case studies were not intended to be formal safety evaluations they were crucial in illustrating important concepts in the safety assessment of novel foods. Therefore, it was through their application in real examples that the concepts and principles were identified. The Working Group proposed a scientific approach to the evaluation of foods derived through modern biotechnology, which is based on a comparison with traditional foods that have a safe history of use. One of the main concepts developed was that of substantial equivalence, and in describing this concept, the Group stressed that this was a principle that had been used in the past (perhaps intuitively) even if it had not been articulated as such.

The concept of substantial equivalence is described as embodying the idea that existing organisms used as food, or as a source of food, can be used as the basis for comparison when assessing the safety of human consumption of a food or food component that has been modified or is new. As previously indicated, in elabourating the concept of substantial equivalence, the OECD Working Group noted that food safety is considered as a reasonable certainty, that no harm will result from intended uses under the anticipated conditions of consumption and that the most practical approach to the determination of safety is to consider whether a genetically modified organism (GMO)-derived food is comparable to an analogous conventional food product.

The concept of substantial equivalence therefore relies on the existing history of safe food use of a conventional food product as a useful consideration in framing the safety assessment of a GMO-derived food by permitting the identification of similarities and differences which can be considered in the assessment. In 1996, participants at an expert FAO/WHO consultation recommended that safety assessment based upon the concept of substantial equivalence be applied in establishing the safety of foods and food components derived from genetically modified organisms.

Assessing substantial equivalence was recognized as not being a safety assessment per se, but a process which establishes that the characteristics and composition of the new GMO-derived food are comparable to those of a familiar, conventional food which has a history of safe consumption. A Joint

FAO/WHO Expert Consultation on Foods Derived from Biotechnology was convened in 2000 to address food safety and nutritional questions regarding foods derived from GM plants, including a review of the scientific basis, application and limitations of the concept of substantial equivalence.

It concluded, based on the current application of substantial equivalence and alternative strategies that this concept contributed to a robust safety assessment framework. Moreover, the Consultation noted that substantial equivalence is a concept used to identify similarities and differences between GM food and a comparator with a history of safe food use which subsequently guides the safety assessment process.

While not a safety assessment per se, the substantial equivalence approach allows the structuring of the safety assessment through characterizing the similarities and identifying the differences which can then be the focus of further consideration. This approach has been referred to as a useful safety standard. It therefore permits inference that the new food under consideration will be no less safe than the conventional food under conditions of similar exposure, consumption patterns and processing practices.

Substantial equivalence is therefore clearly not intended to be a measure of absolute safety, but instead recognizes that while demonstrating absolute safety is an impractical goal, demonstrating that there is reasonable assurance that the GMO-derived product under consideration is no less safe than a conventional food product is an achievable goal. Since it's development, the use of the substantial equivalence concept in the safety assessment of GMO-derived foods has been subject to misinterpretation and criticism.

The comparative nature of the substantial equivalence concept without prescribing the extent of the phenotypic and compositional comparisons has led some to criticize the concept as not being measurable, and therefore inappropriate in safety assessment. This and other criticisms relate, in part, to the mistaken perception that the determination of substantial equivalence was the end point of a safety assessment rather than the starting point.

After several OECD countries had gained experience with safety assessment of GMO-derived foods, an OECD workshop examined the effectiveness of the application of substantial equivalence in safety assessment. This workshop concluded that the substantial equivalence approach provides equal or increased assurance of the safety of foods derived from genetically modified plants, as compared with foods derived through conventional methods.

In 2000, the OECD Task Force for the Safety of Novel Foods and Feeds also reviewed the substantial equivalence concept, its interpretation and its application. While noting that the majority of international guidance documents addressing the safety assessment of genetically modified (GM) plants had interpreted substantial equivalence consistently, this Task Force reported that there were differences in how it was applied and that these differences needed to be resolved. The Codex Ad Hoc Intergovernmental Task Force on Foods Derived from Biotechnology pursued the development of international guidance on the safety assessment of GMO-derived foods. To date, this Task Force has developed proposed draft principles and guidelines for the safety assessment of foods derived from modern biotechnology. These guidelines interpreted the concept of substantial equivalence as a way of structuring the safety assessment, consistent with the FAO/WHO interpretation.

The 'Draft Principles for the Risk Analysis of Foods Derived from Modern Biotechnology' and 'Draft Guideline for the Conduct of Food Safety Assessment of Foods Derived from Recombinant-DNA Plants' have been forwarded by the Codex Task Force to the 25th Session of the Codex Alimentarius Commission for adoption at Step 8 of the Codex procedure. A proposed draft guideline for the food safety assessment of foods produced using recombinant-DNA organisms continues to be developed. Substantial equivalence: the application

A conclusion from a 1995 WHO workshop on 'Application of the Principles of Substantial Equivalence to the Safety Evaluation of Foods or Food Components from Plants Derived by Modern Biotechnology' summarized the relationship between substantial equivalence and safety assessment as follows: Establishment of substantial equivalence is not a traditional safety assessment in itself, but a dynamic, analytical exercise in the assessment of the relative safety of a new food or food component to an existing food or food component. The application of the substantial equivalence concept is a key step in structuring the safety assessment of GMO-derived foods. Through appropriate analysis, the new food is compared phenotypically and compositionally to its conventional counterpart with a history of safe use. In this compositional comparison, the characteristics, including levels of key nutrients and toxicants, are considered relative to those of the conventional counterpart taking into account the natural variation for such characteristics.

When a genetic modification results in the insertion of a specific trait, the comparative analysis of the new food will identify both the intended effect (inserted trait) and the potential unintended effects of the modification. Further assessment can then focus on new or altered characteristics for which no history of safe use can be established.

One of the important benefits of applying the substantial equivalence concept is that it provides flexibility which can be a powerful tool in terms of food safety assessment. The comparative approach to structuring the safety assessment can be applied at several potential levels along the food continuum (i.e. harvested primary food material or unprocessed food product, individual processed fractions, or final food product or ingredient).

While from a practical point of view, the compositional comparison should typically be applied at the level of the unprocessed food product, the flexibility of the concept permits the determination to be targeted to the most appropriate level based upon the nature of the product under consideration. Where multiple fractions from a single source are destined to different food products, the comparative approach might be targeted at the level of the unprocessed food product in order to permit the safety assessment to apply to all derived fractions (e.g. for soybean, where multiple fractions are used widely in foods, assessment at the level of the seed is appropriate).

However, where a single fraction of a particular raw material is used as human food, then the comparison can focus at the level of the single fraction, thereby simplifying the safety assessment process (e.g. for canola, where the processed oil is the only fraction consumed by humans, safety assessment may appropriately be focused on comparison of the oil composition of the novel variety with traditional canola oil composition).

Application of the substantial equivalence concept in the safety assessment of a GMO-derived food depends on the identification of an appropriate comparator with an acceptable history of safe food use. It also requires that sufficient analytical data be available in the literature or be generated through analysis to permit an effective comparison. These requirements present a key limitation of the substantial equivalence concept since the consideration of similarities can only provide assurance of safety relative to those components assessed for the particular comparator. The choice of the comparator is therefore crucial to the effective application of substantial equivalence in establishing the safety of a GMO-derived food.

An appropriate comparator must have a well-documented history of use. If adverse effects have been associated with the particular food type, specific components of the food which are considered to be causative of those adverse effects should be described and well characterized in order to permit effective comparison. A joint FAO/WHO Expert Consultation on Biotechnology and Food Safety considered the application of substantial equivalence in the safety assessment of GMO-derived foods.

The consultation recommended that applicacation of the substantial equivalence concept entail consideration of the molecular characterization of the new food source; phenotypic characterization of the new food source in comparison to an appropriate comparator already in the food supply; and the compositional analysis of the new food source or the specific food product in comparison to the selected comparator. The guidance further elabourates on compositional comparison by highlighting what information should provide sufficient information to permit effective comparison.

The recommended focus of the compositional comparison is on analytical comparison of those components identified in the food source in question which are nutrients which provide a substantial impact in the overall diets (key nutrients) and toxicologically significant compounds known to be inherently present in the species (key toxicants). In addition, there is recognition that additional components might be identified for analysis based upon the molecular and phenotypic characterization and the nature of the genetic modification. In addition to key nutrients and toxicants, additional parameters may be appropriate for assessing the potential for unintended effects of a genetic modification.

When modifications are directed at metabolic pathways of key macro or micro nutrients, the possibility of an impact on nutritional value is increased, particularly if that food is a major dietary source of the nutrient affected. The potential for unintended effects would be determined, in part, by the nature of the intended alteration (e.g. consideration of fatty acid profile if an enzyme involved in fatty acid metabolism is introduced) and the data from molecular and phenotypic characterization. The consideration of key nutrients and key toxicants in the comparison which is essential to applying a substantial equivalence in safety assessments also introduces another limitation of the concept. The nature of the comparative approach with respect to nutrients limits its universality since the relevance of nutrients in a particular crop are dependent on consumption patterns which might vary from region to region. Where differences in consumption exist for a particular crop, these must be considered in the identification of the key nutrients for assessment.

This is particularly true for crops which form a significant portion in the diet in a particular region. The comparative approach to assessment can be applied in each region, but conclusions for one region will not automatically hold for another region if there are significant differences in consumption patterns and processing practices. Another potential regional limitation is related to the application of the concept of substantial equivalence as opposed to an inherent limitation of the concept per se.

It may be difficult, particularly in developing countries, to apply the concept to assess the safety of foods where adequate nutritional databases are not available for a given population. A safety assessment using the substantial equivalence approach does not demonstrate that a GMO-derived product is identical to its conventional comparator since the compositional comparison does not take into account all components.

However, application of the guidance provides assurance that the comparison has considered those components most likely to be relevant to the safety of the product as it is expected to be consumed in a particular region. Recognizing the importance of there being compositional data available for applying substantial equivalence, the OECD Task Force for the Safety Assessment of Novel Foods and Feeds has focused on the development of science-based consensus documents containing information on the nutrients, anti-nutrients or toxicants, product use and other data relevant to the assessment.

The OECD have published such documents for potatoes, sugar beet, soybean and low erucic acid rapeseed (canola). Such guidance will add to the understanding of appropriate parameters for a comparative assessment of composition while recognizing that additional parameters may be relevant to the safety assessment dependent on differences in consumption pattern. The international guidance developed by the OECD and FAO/WHO has been practically applied to the safety assessment of GMO-derived food products in several countries. In order to facilitate such assessments, specific guidance documents which embrace the substantial equivalence concept have been published. The majorities of these currently address the safety assessment of genetically modified plants and have consistently interpreted the concept of substantial equivalence. Internationally, these guidance documents have been applied to the assessment of a significant number of GMO-derived plant products over a period of more than eight years, demonstrating that the concept of substantial equivalence can be applied effectively in the safety assessment of novel foods.

IMPACTS OF FOOD-CROP IMPROVEMENT TECHNOLOGY

Substantial empirical evidence exists on the production, productivity, income and human welfare impacts of modern agricultural science and the international flow of modern varieties of food crops. Evenson and Gollin provide detailed information on the extent of adoption and impact of modern variety use for all the major food crops. The adoption of modern varieties (averaged across all crops) increased rapidly during the two decades of the Green Revolution, and even more rapidly in the following decades, from 9 per cent in 1970 to 29 per cent in 1980, 46 per cent in 1990 and 63 per cent by 1998. Moreover, in many areas and in many crops, first-generation modern varieties have been replaced by second- and third-generation modern varieties.

Much of the increase in agricultural output over the past 40 years has come from an increase in yield per hectare rather than an expansion of area under cultivation. For instance, FAO data indicate that for all developing countries, wheat yields rose 208 per cent from 1960 to 2000; rice yields rose 109 per cent; maize yields rose 157 per cent; potato yields rose 78 per cent and cassava yields rose 36 per cent. Trends in total factor productivity are consistent with partial productivity measures, such as rate of yield growth.

The returns to investments in high-yielding modern germplasm have been measured in great detail by several economists over the last few decades. Several recent reports have reviewed and analysed the data from hundreds of studies conducted over the last 30 years that calculated the social rates of return to investments in agricultural research. These studies examined investments by national and international public-sector institutions in Africa, Asia, Latin America and the Organisation for Economic Co-operation and Development (OECD) countries as well as by the private sector. Although these studies were carried out using a variety of different methods, they showed considerable consistency. The average social rate of return to public investment in agricultural research reported in these studies is in the region of 40-50 per cent. Private-sector research was also found to generate similar rates of social returns.

The primary effect of agricultural research on the non-farm poor, as well as on the rural poor who are net purchasers of food, is through lower food prices. The wide-spread adoption of modern seed-fertilizer technology led to a significant shift in the food supply function, increasing output and contributing to a fall in real food prices:

- The effect of agricultural research on improving the purchasing power of the poor - both by raising their incomes and by lowering the prices of staple food products - is probably the major source of nutritional gains associated with agricultural research. Only the poor go hungry. Because a relatively high proportion of any income gains made by the poor is spent on food, the income effects of research-induced supply shifts can have major nutritional implications, particularly if those shifts result from technologies aimed at the poorest producers.

Studies by economists have provided empirical support for the proposition that growth in the agriculture sector has economy-wide effects. Hayami *et al.* illustrated at the village level that rapid growth in rice production stimulated demand and prices for land, labour and non-agricultural goods and services. For sector-level validation of the proposition that agriculture does indeed act as an engine of overall economic growth.

Once modern varieties have been adopted, the next set of technologies that makes a significant difference in reducing production costs includes machinery, land management practices (often in association with herbicide use), fertilizer use, integrated pest management and (most recently) improved water management practices. Although many Green Revolution technologies were developed and extended in package form (*e.g.* new plant varieties plus recommended fertilizer, pesticide and herbicide rates, along with water control measures), many components of these technologies were taken up in a piecemeal, often stepwise manner. The sequence of adoption is determined by factor scarcities and the potential cost savings achieved. Herdt provided a detailed assessment of the sequential adoption of crop management technologies for rice in the Philippines. Traxler and Byerlee provided similar evidence on the sequential adoption of crop management technologies for wheat in Sonora, northwestern Mexico.

Although the favourable, high-potential environments gained the most from the Green Revolution in terms of productivity growth, the less favourable environments also benefited through technology spillovers and through labour migration to more productive environments. According to David and Otsuka (1994), wage equalization across favourable and unfavourable environments was one of the primary means of redistributing the gains of technological change. Renkow found similar results for wheat grown in high- and low-potential environments in Pakistan. Byerlee and Moya, in their global assessment of the adoption of modern varieties of wheat, found that over time the adoption of modern varieties in unfavourable environments caught up with those in more favourable environments, particularly when germplasm developed for high-potential environments was further adapted to the more marginal environments. In the case of wheat, the rate of growth in yield potential in drought-prone environments was around 2.5 per cent per year during the 1980s and 1990s. Initially, the growth in yield potential for the marginal environments came from technological spillovers as varieties bred for the high-potential environments were adapted to the marginal environments. During the 1990s, however, further gains in yield potential came from breeding efforts targeted specifically at the marginal environments.

Even while highlighting the yield breakthrough in wheat, the Government had also launched a massive programme to develop and spread high-yielding varieties for rice, maize, sorghum and pearl millet. These programmes were the drivers of the "Green Revolution" in India, which permitted striking advances in production and productivity without increasing cultivated area.

Because these high-yielding varieties require inputs such as fertilizer and irrigation water, social scientists criticized the Green Revolution technologies for not being resource neutral. Environmentalists attacked the Green Revolution because of potential damage to long-term productivity as a result of excessive use of pesticides and fertilizers and monocropping. Despite the success of the Green Revolution in raising millions of people out of misery, the incidence of poverty, endemic hunger, communicable diseases, infant and maternal mortality rates, low birth-weight children, stunting and illiteracy remain high.

The concerns of social scientists and ecologists and the remaining urgent problems of poverty and hunger led to my developing the concept of an "evergreen revolution" to stress the need for enhancing crop productivity in perpetuity without associated ecological or social harm. An evergreen revolution can be achieved only if we pay attention to pathways that can help to achieve revolutionary progress in enhancing productivity, quality and value-addition under conditions of diminishing per capita arable land and irrigation water availability, expanding biotic and abiotic stresses, and fast-changing consumer and market preferences. This will require mobilizing the best in both traditional wisdom and technologies and frontier science. Among the frontier technologies relevant to the next stage in our agricultural revolution, the foremost is biotechnology.

The apprehensions relating to molecular genetics and genetic engineering fall under the following broad categories: the science itself, the control of the science, access to the science, environmental concerns, and human and animal health. A disaggregated approach to the study of these issues will be important for a rigorous analysis of risks and benefits. Dealing with these issues in a composite manner for all applications of genetic engineering will result in inappropriately broad conclusions, such as the general condemnation of GMOs expressed by non-governmental organizations (NGOs) at the World Food Summit: *five years later* held in Rome in 2002.

The benefits of molecular breeding techniques such as the use of molecular markers and undertaking precision breeding for specific characters through recombinant DNA technology are immense. The work already performed in India has revealed the potential for breeding new GM varieties possessing tolerance to salinity, drought and some major pests and diseases, together with improved nutritive quality. A new era of integrated Mendelian and molecular breeding has begun.

An evergreen revolution will blend these frontier technologies with the ecological prudence of traditional communities to create technologies that are based on integrated natural resource management and that are location specific because they are developed through participatory experimentation with farm families.

This is the only way we can face the challenges of the future, particularly in the context of the growing water scarcity and the urgent need to step up productivity in semi-arid and dry farming areas. Accelerated agricultural progress is the best safety net against hunger and poverty, because in most developing countries over 70 per cent of the population depend on agriculture for their livelihood. Denying ourselves the power of the new genetics will be doing a great disservice both to resource-poor farming families and to the building of a sustainable national food and nutrition system.

THE GENE REVOLUTION: A CHANGING PARADIGM FOR AGRICULTURAL R&D

In the 1960s, 1970s and 1980s, private-sector investment in plant improvement research was limited, particularly in the developing world, owing to the lack of effective mechanisms for proprietary protection of the improved products.

This situation changed in the 1990s with the emergence of hybrids for cross-pollinated crops such as maize. The economic viability of hybrids led to a budding seed industry in the developing world, started by transnational companies from the developed world and followed by the development of national companies. Despite the rapid growth of the seed industry in developing countries, its activity has been limited to date, leaving many markets underserved.

The incentives for private-sector agricultural research increased further when the United States and other industrialized countries permitted the patenting of artificially constructed genes and genetically modified plants. These national protections were strengthened by the 1995 Agreement on Trade

Related Aspects of Intellectual Property Rights (TRIPS) of the World Trade Organization (WTO), which obliges WTO members to provide patent protection for biotechnology inventions (products or processes) and protection for plant varieties either through patents or a *sui generis* system. These proprietary protections provided the incentives for private sector entry in agricultural biotechnology research.

The large transnational agrochemical companies were the early investors in the development of transgenic crops, although much of the basic scientific research that paved the way was conducted by the public sector and made available to private companies through exclusive licences. One of the reasons why agrochemical companies moved into transgenic crop research and development was that they foresaw a declining market for pesticides and were looking for new products.

The chemical companies moved quickly into plant improvement by purchasing existing seed companies, first in industrialized countries and then in the developing world. These mergers between national seed companies and multinational corporations made economic sense because the two specialize in different aspects of the seed variety development and delivery process. This process is a continuum that starts upstream with generating knowledge on useful genes (genomics) and engineering transgenic plants and then moves downstream to the more adaptive process of backcrossing the transgenes into commercial lines and delivering the seed to farmers.

The products from upstream activities have worldwide applicability across several crops and agro-ecological environments. By contrast, genetically modified crops and varieties are typically applicable to specific agro-ecological niches. In other words, spillover benefits and scale economies decline in the move to the more adaptive end of the continuum.

Similarly, research costs and research sophistication decline in the progression towards downstream activities. Thus, a clear division of responsibilities in the development and delivery of biotechnology products has emerged, with the transnational firm providing the upstream biotechnology research and the local firm providing crop varieties with commercially desirable agronomic backgrounds.

The options available for capturing the spillovers from global corporations are less clear for public research systems. Public-sector research programmes are generally established to conform to state or national political boundaries, and direct country-to-country transfer of technologies has been limited. Strict adherence to political domains severely curtails spillover benefits of technological innovations across similar agroclimatic zones. The operation of the CGIAR germplasm exchange system has mitigated the problem for several important crops, but it is not clear whether the system will work for biotechnology products and transgenic crops, given the proprietary nature of the technology.

GOODS

Public goods are those that generate benefits for society beyond the private returns that can be captured by the person who created them. These benefits are sometimes called spillovers. Public goods are non-rival and non-excludable. Non-rivalry implies that the good is equally available to all, *i.e.* consumption by one person does not reduce the amount that is available for others to consume. Non-excludability means that people who do not pay for the product cannot be prevented from using it. These characteristics mean that private innovators cannot capture the full social benefit of their creation unless some means can be found to prevent unauthorized use. Because private firms cannot profit fully from research that produces public goods, they will not invest in a socially optimal level of research.

Much of the output of agricultural research, including biotechnology research, has one or both of the characteristics of a public good. For example, any scientist can use knowledge about the structure of the rice genome without reducing the amount of knowledge available to other scientists, and once that knowledge is published in an academic journal or on the Web, it is difficult to exclude other people from using it. A transgenic plant variety, on the other hand, may have public good characteristics to some degree (*e.g.* it is difficult to exclude unauthorized users completely) but it is not a pure public good because seeds can be used up and unauthorized use can be at least partially prevented.

There are two ways to prevent the unauthorized use of plant varieties — biological and legal. Hybrid seeds can be saved, reproduced and replanted but only at a significant loss in yield and quality, so hybridization provides a biological protection for the breeder's innovation. Genetic use-restriction technologies are another form of biological intellectual property protection that has been proposed for transgenic crops. These technologies would produce sterile seeds or seeds that require the application of a special chemical to activate the innovative trait. Public opposition to the sterile-seed approach has led the private company Monsanto to abandon its development. Legal protection such as patents, trademarks and contracts can also be used to protect intellectual property, but these methods usually provide incomplete protection.

BIOTECHNOLOGY RESEARCH INVESTMENTS

To understand the magnitude of private-sector investment in agricultural biotechnology research today, one need only look at its annual research budget relative to public research targeted at developing country agriculture. The world's top ten transnational bioscience corporations' collective annual expenditure on agricultural biotechnology research and development is nearly \$3 billion. By comparison, the CGIAR, which is the largest international public-sector supplier of agricultural technologies, has a total annual budget of less than $300 million for plant improvement research and development. The largest public-sector agricultural research programmes in the developing world - those of Brazil, China and India - have annual budgets of less than half a billion dollars each.

Looking at agricultural biotechnology research expenditures reveals a sharp dichotomy between developed and developing countries (Table). Developed countries spend four times as much as developing countries on public-sector biotechnology research, even when all sources of public funds - national, donor and CGIAR centres - are counted for developing countries. Few developing countries or international public-sector institutions have the resources to create an independent source of biotechnology innovations.

Comprehensive data on private-sector biotechnology research in developing countries are not available, although most research appears to be carried out by transnational companies conducting trials of their transgenic varieties. Some work is being done by local research institutes (*e.g.* local private sugar-cane research institutes have fairly large biotechnology research programmes in Brazil and South Africa), whereas in India several local seed companies biotechnology research programmes. The total investment of these private efforts is unknown but it is undoubtedly less than the public sector is investing in biotechnology research in developing countries.

TRANSFER OF DNA FROM GMO IN FOODSTUFFS

There is potential for the presence of live GMOs in foodstuffs such as yoghurt and fermented sausage produced with starter cultures containing genetically modified organisms or perishable foods treated with protective cultures of GM organisms. The environment within living food may permit gene transfer by conjugation or transduction and this has been seen at high frequency in complex food matrices.

Free recombinant DNA may also be present in food products resulting from the killing of GMMs during food processing. The human gut microflora is extremely complex, being made up of more than 500 different species of bacteria and climax microbial populations reaching up to 1011 CFU/g luminal contents in the colon. The high density and species diversity of bacteria present in the colon provides ideal conditions for the transfer of DNA between bacteria.

However, the diversity of the human gut microflora also presents significant problems in monitoring the DNA transfer events between bacteria and, in particular, monitoring the fate of recombinant DNA, which may be ingested in food. It has been determined that only a fraction of the bacteria present in the human gut microflora may be cultured under labouratory conditions and characterized.

Many species found in the human gut microflora defy culture by standard microbiological techniques, and selective agars and growth conditions exist for only a small proportion of the predominant species present. Much progress as been made in recent years in the development of culture-independent methods for identifying bacteria present in natural environments.

Using the phylogenetic information present in 16S rRNA and fluorescent in situ hybridization techniques, it is now possible to identify bacteria in situ in natural environmental samples including human faeces. However, such techniques tell us little about the genetic capabilities of bacteria and by themselves do not allow us to monitor the distribution or indeed dissemination of specific genetic determinants amongst natural assemblages of bacteria.

Thus, studies on the transfer of DNA between members of the human gut microflora and from GMOs to members of the human gut microflora have employed specific marker systems borne on the mobile genetic elements or on recombinant DNA. Traditional marker systems used in pure culture microbiology such as the lac operon or X-gal are of little use when employed to monitor DNA transfer events in complex assemblages of gut microorganisms, because of the widespread occurrence of such genes amongst the mammalian gut microflora.

Antibiotic-resistance determinants, when shown to be absent from background bacteria, have been widely employed in labouratory-based studies. However, such marker systems cannot be used in studies with human volunteers because of the emergence of antibiotic-resistance determinants in pathogenic strains. Novel marker systems such as green fluorescent protein are now being employed in labouratory studies to monitor the persistence of GMOs in the human gut microflora and DNA transfer events between bacteria. Another hurdle to monitoring the dissemination of recombinant DNA in complex microbial consortia and investigating the factors governing DNA transfer in natural environments is the non-expression of marker genes in novel bacterial hosts.

Marker genes transferred between bacteria may not always be expressed in the recipient cell. Thus not all DNA transfer events will be detected by existing marker systems. One way around this problem is to use a specific plasmid transfer system comprising a donor and recipient bacterium in which the marker gene is known to be expressed, and then examining the factors governing DNA transfer between these two strains in models of the ecosystem under investigation.

Recent advances in molecular biology offer some hope that more reliable means of monitoring the dissemination of genetic determinants amongst diverse bacteria may be developed. Fluorescent in situ hybridization is now being employed to enumerate bacteria in natural assemblages by hybridizing 16S rRNA targets to fluorescently labelled DNA probes and visualization with fluorescent microscopy. Similarly, it is now possible to visualize PCR products within bacterial cells fixed on microscope slides. By combining these techniques it may soon be possible to visualize concomitantly, fluorescently labelled PCR products of single-copy genes and labelled 16S rRNA sequences within a single cell in environmental samples. To enable us to ask specific questions about the human gut microflora, in vitro and in vivo models of the human gut have been developed.

In vitro models generally employ anaerobic chemostats of varying degrees of complexity or short-term batch cultures of human faeces or luminal samples extracted from animals. They allow generation of data on the effect of specific biotic and abiotic environmental parameters on the gut microflora, which may otherwise be unavailable due to inaccessibility of the target environment in the gastrointestinal tract or for ethical reasons.

Freter *et al.* studied the transfer of plasmid DNA between E. coli strains using a single-stage continuous flow culture model of the mouse gut microflora. This in vitro model of the mouse caecal microflora simulated microbial interactions observed in the mouse gut. Data generated from the in vitro model also allowed validation of mathematical models of plasmid transfer between E. coli strains in the presence of the complex mouse gut microflora. Rang *et al.* employed short-term batch cultures of luminal extracts from the mouse gastrointestinal tract to monitor transfer of plasmid RP1 between E. coli strains. More complex in vitro models of the human gastrointestinal tract developed recently have yet to be employed to monitoring DNA transfer events in the human gut microflora.

In vitro models allow us to examine interactions between specific members of the gut microflora or between the whole human faecal microflora and introduced GMOs and non-GMOs under the physiological conditions of the human gut.

They provide a vital first stage in the biosafety assessment of GMOs intended for use in human food. However, such models do not take into consideration the biological complexity of the mammalian gut ecosystem. In vitro models may not simulate the biological surfaces or the input of endogenous substrates and secretions of the human immune system, which may play a vital role both in the maintenance and establishment of the gut microflora, and the frequency of DNA transfer events.

Gnotobiotic and germ-free animals have been employed as in vivo models of the human gastrointestinal ecosystem for some years. Gnotobiotic technology provides a useful tool in studying the interactions between specific members of the human gut microflora, e.g. the role played by specific members of the microflora in colonization, resistance to pathogenic bacteria, bacterial pathogenesis, and metabolism of dietary constituents or xenobiotic compounds.

Transfer of plasmid DNA between specific members of the human gut microflora has also been studied using di- or poly-associated animals. Conventional animals may not provide information of direct relevance to the human gut ecosystem because of the major differences in the composition of the gut microflora between animals and man. To circumvent these problems, germ-free animals have been associated with groups of bacteria from the human gut microflora or the whole human gut microflora.

Such human flora-associated (HFA) animals provide an essential tool in studies of human gut microbiology in that they reflect the complexity of the human gut ecosystem in terms of microbial species and numbers, and they provide some of the biotic parameters provided by the mammalian mucosa, immune system and digestive functions.

Such in vivo models provide an extremely useful tool in the generation of data on the safety of GMOs in food until a greater degree of confidence in the safety of such products is attained and human volunteer studies can ethically be considered.

Factors Affecting Transfer of Plasmid DNA from Bacteria in vivo

Very little is known about the extent to which transformation and transduction contribute to DNA transfer between bacteria in the human gastrointestinal microflora. Only one study on bacterial transformation in fresh human saliva samples has been presented in the literature. No reports of transformation or transduction in lower regions of the human gastrointestinal tract were found after exhaustive search of the literature.

This is probably due to the fact that conjugation has been viewed as the dominant contributor to DNA transfer between bacteria in natural environments, because of the ubiquitous occurrence of conjugative elements amongst diverse bacterial species. Thus most studies looking at DNA transfer in the gut have concentrated on conjugation.

Earlier, it was proposed that the frequency of plasmid transfer in a given environment will be a function of the interaction between environmental factors and the bacterial populations involved, characteristics of the plasmid itself and the mode of conjugation employed by the plasmid. Some of these parameters have been studied in vivo in animal models.

Effect of Antibiotics on Plasmid Transfer

A number of early investigators observed the transfer of plasmid DNA between bacteria in the gastrointestinal tract of human volunteers. However, many of these studies were conducted with antibiotic-resistance determinants in E. coli strains and in the presence of antibiotic selective pressure. In the light of the recent emergence of pathogenic bacteria with multiple antibiotic resistance determinants it is no longer considered ethically justifiable to carry out such investigations in human volunteers.

Ingestion of antibiotics leads to perturbation of the natural homeostasis of the human gut microflora and presents a selective pressure for the evolution of antibiotic resistance strains through mutation and transfer of resistance determinants. The human gut microflora is commonly exposed to antibiotics, both through clinical practice and use of antibiotics in animal husbandry.

Similarly, antibiotic-resistant bacteria have been isolated from a wide range of human foodstuffs. Duval-Iflah *et al.* observed the transfer of an R plasmid from Serratia liquifaciens to an E. coli recipient strain, naturally present in the human microflora of the HFA mice. The donor S. liquifaciens strain became established in the HFA mice at between 106 and 107 CFU/g faeces. Transfer was observed 12 days after inoculation with the donor strain and their numbers ranged from between 103 and 105 CFU/g faeces. The number of transconjugants greatly increased after addition of selective antibiotic in drinking water. Morelli *et al.* showed that a subtherapeutic dose of selective antibiotic (erythromycin at 10 μg/ml drinking water) increased the level of pAMâ1 transfer from L. reuteri to Ent. faecalis from 1 × 10-7 to 1 × 10-4 transconjugants per donor in anexic mice (i.e. gnotobiotic mice associated with donor and recipient strains).

Although selective antibiotic pressure previously has been employed to demonstrate transfer of genetic elements between bacteria in animal models and in humans, the observation that subtherapeutic doses of selective antibiotic increase the rate of plasmid transfer has important implications for both the design in GMOs and the use or misuse of antibiotics in clinical practice and agriculture.

Persistence of GMOs and Their Recombinant DNA in vivo

The ability of GMOs, which may be ingested in a viable form in fermented foods, to persist in the human gastrointestinal tract will greatly determine the extent to which recombinant DNA may be transferred to the human gut microflora.

Similarly the stability of recombinant DNA within GMOs in vivo is of major interest when designing GMOs for use in food or in the human gut itself. Stable maintenance of recombinant DNA will greatly determine the ability of a GMO to effectively carry out the task for which it was designed. On the other hand, a certain degree of recombinant DNA instability may limit the risks posed by recombinant DNA persisting in non-target ecosystems.

Gruzza *et al.* investigated the colonization potential and recombinant DNA stability of genetically modified L. lactis strains in germ-free mice. All L. lactis constructs reverted to plasmid-free derivatives in the digestive tract of anexic mice and plasmid-free derivatives became dominant over plasmid-bearing parental strains. Two high copy number, non-self transmissible plasmids exerted strong ecological disadvantage on plasmid-bearing L. lactis strains and were quickly lost from the digestive tracts of the anexic mice.

Transfer of the conjugative plasmid pIL205 was observed between an Ent. faecalis strain colonizing the digestive tract of anexic mice and an L. lactis donor strain. Both the L. lactis donor strain and Ent. faecalis (pIL205) transconjugants were lost from the anexic mice 10 days after dosing with the donor strain. Thus it appears that carriage of recombinant DNA on non-conjugative plasmids and indeed conjugative plasmid DNA, imposed considerable ecological disadvantages on plasmid-bearing strains in anexic animals. Schlundt *et al.* investigated transfer of pAMâ1 between L. lactis donor and recipient strains in gnotobiotic and conventional rats. In the conventional animals, both donor and recipient strain were quickly lost from the system and no transconjugants were recovered.

In gnotobiotic animals both donor and recipient strain colonized the gastrointestinal tract. Tranconjugants were recovered within a few days of dosing and maintained at between 103 and 105 CFU less than the recipient strain in faecal samples. Upon sampling of luminal contents of the jejunum, caecum and colon it was observed that carriage of the plasmid seemed to endow a competitive advantage to transconjugants in the small intestine but not in the caecum.

The population of transconjugants was maintained at about 104 CFU/g throughout the intestine but numbers of recipient strain varied from between 104 and 105 CFU/g jejunum contents to 108-109 CFU/g caecal and colonic contents. Clearly the ability of plasmid DNA either to impose ecological advantage or disadvantage on the bacterial host will depend on the plasmid and bacterium involved, and may very well vary considerably between different environmental habitats or microhabitats.

Effect of Microbial Interactions on DNA Transfer

The human gut microflora exerts considerable colonization resistance on extraneous microorganisms entering the colonic environment. Limiting the persistence of a GMO in a given environment will also limit its ability to undergo DNA transfer with the microflora. Gruzza *et al.* investigated the ability of members of the human gut microflora to limit the colonization ability of and DNA transfer from, genetically modified Lactococcus lactis strains, similar to those likely to be used in the food industry.

L. lactis strains bearing the conjugative plasmid pIL205 and/or a non-conjugative but a pIL205 mobilizable plasmid pIL253 were used. Gnotobiotic mice were associated with four strictly anaerobic bacteria commonly found in the human gut microflora, i.e. Bacteroides sp., Bifidobacterium sp., Peptostreptococcus sp. and Ent. faecalis.

No transfer of plasmid DNA was observed from L. lactis donor strain to Bacteroides sp., Bifidobacterium sp. or the Peptostreptococcus sp. However, upon addition of Ent. faecalis to the polyassociated mice, Ent. faecalis transconjugants were observed despite the competitive exclusion imposed on the L. lactis donor strain by the colonizing Ent. faecalis strain. Gene transfer from the L. lactis strains was also investigated in HFA mice. HFA mice were dosed with L. lactis donor strain, bearing pIL205 and pIL253, seven days after introduction of the human faecal microflora. From an initial inoculum of 8 × 108 CFU, no L. lactis donor cells were recovered from the faecal samples 24 hours after dosing (detection limit 102 CFU/g faeces). However, 21 hours after introduction of the donor strain, transconjugant facultatively anaerobic streptococci were recovered resistant to antibiotic markers borne on each of the two plasmids.

The number of transconjugants increased until day 15 after dosing. Thus, although rather quickly eliminated from the HFA mice, the L. lactis donor strain was able to transfer its plasmid DNA (both the conjugative pIL205 and the non-conjugative but mobilizable pIL253) to members of the human gut microflora in HFA mice.

Brockmann *et al.* investigated the transfer of plasmid pAMâ1 and pLMP1 (bearing a genetically modified proteinase gene) from L. lactis donor strains to a proteinase-deficient recipient strain in gnotobiotic and conventional rats.

No transfer of the proteinase gene to the recipient L. lactis strain was observed in anexic or conventional rats. Plasmid pAMâ1 was transferred between L. lactis donor and recipient strains in anexic animals but not in conventional rats where both donor and recipient were quickly lost from faecal samples. Transfer of pAMâ1 from the L. lactis donor strain to Ent. faecalis (or a close relative) was observed in one conventional rat.

Using an in vitro single stage continuous flow culture (CFC) of human faecal microflora and in vivo human flora-associated (HFA) rat model of the human colonic microflora we have shown that the human gut microflora exerts a considerable barrier to persistence of L. lactis strains in the colonic ecosystem. A plasmid-free L. lactis strain MG1614 was washed out from the in vitro model at a rate greater than expected for non-growing cells.

Similarly, L. lactis MG1614 introduced into HFA rats was eliminated from the HFA rats within eight days. By continuously dosing HFA rats with L. lactis MG1614 in the drinking water we succeeded in maintaining a stable population of L. lactis MG1614 in the rats. It was hoped that this strain implanted in the microflora of HFA rats would act as a recipient in plasmid transfer studies with a closely related L. lactis strain.

Transconjugants were maintained at between 104 and 105 CFU/g faeces for 27 days. Thus despite the fact the human gut microflora exerts such a drastic barrier effect on the persistence of L. lactis strains in models of the human colonic microflora, transfer of plasmid DNA from an ingested L. lactis donor strain to Enterococcus spp. present in the microflora was observed.

Design of GMOs and Colonization of Gut by GMOs

GMOs considered for release into the environment in a viable form will employ several measures to limit the ability of recombinant DNA to be transferred to indigenous bacteria in the target environment. Recombinant DNA may be borne on non-conjugative vectors, on the chromosome of the host strain or by the use of suicidal genetic elements. However, inbuilt safeguards which prevent the transfer of recombinant DNA under labouratory conditions cannot be relied upon to prevent transfer of recombinant DNA in the natural environment. Thus studies employing in vivo models of the human gastrointestinal tract have been conducted with such GMOs in order to determine their ability to persist in the digestive tract and transfer DNA. Gruzza *et al.* investigated the ability of genetically modified L. lactis strains to colonize gnotobiotic mice.

The genetically modified L. lactis strains were constructed with antibiotic-resistance marker genes inserted on different replicons, chloramphenicol resistance was inserted into a self-transmissible plasmid pIL205, erythromycin resistance was inserted into non-self transmissible plasmids pIL252 (low copy number) and pIL253 (high copy number) and a second erythromycin-resistance determinant was implanted into the L. lactis chromosome.

All strains carried a naturally occurring pIL9 plasmid encoding the machinery for lactose fermentation. It was observed that anexic mice were colonized with the recombinant strains or their plasmid-free derivatives, with plasmid-free derivatives becoming dominant in all cases. Plasmids pIL9 and pIL205 were lost from donor strains but plasmid-bearing parental strains and plasmid-free derivatives were maintained at co-dominant levels, suggesting equilibrium between plasmid loss and conjugation in the mouse gastrointestinal tract.

L. lactis strains bearing the low-copy number non-self transmissible plasmid pIL252 rapidly lost their plasmid complement in the gut of anexic mice, probably due to high segregational instability of this plasmid observed in vitro. L. lactis strains bearing pIL253, the high-copy number, non-self transmissible plasmid colonized the digestive tract of anexic mice but only became dominant when plasmid pIL205 was not present in the same cell, indicating a degree of incompatibility between the two plasmids both plasmids belong to the same incompatibility group. The erythromycin-resistance determinant borne chromosomally was maintained stably in the digestive tract of anexic mice. The authors concluded that in vitro determinations of recombinant DNA stability may highlight if a particular insert is extremely unstable but does not give reliable data on the in vivo stability of the construct.

This, highlights the need for in vivo studies both on the stability of recombinant DNA for efficient GMO function and the effect instability has on the persistence of recombinant DNA in the digestive tract, which is a major factor governing the extent to which DNA transfer between ingested GMOs and members of the human gut microflora occurs.

Duval-Iflah *et al.* investigated the mobilization of pBR-derived recombinant plasmids from donor E. coli K12 strains to E. coli strains of human origin in vitro and in the digestive tract of anexic and gnotobiotic mice associated with the human gut microflora. The genetically modified plasmids used were designed to be non-conjugative (mob-, tra-) and either mobilizable or non-mobilizable. E. coli strains isolated from human faeces were screened for their ability to mobilize the oriT- and oriT+ plasmids. In anexic mice, associated with donor and recipient E. coli strains, mobilization of an oriT+ pBR derivative by trans-complementation with the gene products of tra (encoded by conjugative plasmid R388) and mob was observed. Such mobilization was only observed sporadically with one E. coli strain of human origin in di-associated mice and not at all in HFA mice. It was determined that about 50% of E. coli strains of human origin were able to mobilize oriT- pBR derivatives in in vitro matings but not in anexic mice or HFA animals.

Dietary Factors Affecting DNA Transfer

The ecological significance of the metabolic burden imposed on a bacterium by plasmid carriage will be greatly determined by the availability of substrates and nutrients to the cell. Similarly, densities of donor and recipient cells have been shown to affect conjugation frequencies in vitro. In our studies, transfer of plasmid pAMâ1 from an L. lactis donor strain to an Enterococcus spp. recipient strain naturally present in the microflora of HFA rats was found to be strongly affected by dose of donor strain employed.

Significant numbers of Enterococcus spp. (pAMâ1) transconjugants were recovered only at a donor cell concentration of 109 CFU/ml or above. The diet fed to the HFA rats also had an effect on transfer of pAMâ1 from the L. lactis donor strain to the Enterococcus spp. in vivo. A high-fat, low-fibre synthetic diet (similar to human high-fat diets) reduced the numbers of Enterococcus spp. (pAMâ1) transconjugants recovered from the faeces of HFA rats compared with the commercial rodent chow (higher levels of complex carbohydrate and fibre).

The presence of increased substrate availability in the caecum of HFA rats fed the rodent chow or increased surfaces provided by the higher fibre content of the rodent chow may be involved in the higher numbers of Enterococcus spp. (pAMâ1) transconjugants recovered from HFA rats fed this diet. Duval-Iflah *et al.* looked at the effect of milk fermented with Lactobacillus bulgaricus and Streptococcus thermophilus on the transfer and mobilization of plasmids between E. coli strains in the gut of anexic mice.

Plasmid transfer studies were conducted in gnotobiotic mice associated with E. coli K12 donor strain and an E. coli recipient strain PG1 of human origin. The affect of milk fermented with L. bulgaricus, S. thermophilus or with both lactic acid bacteria on transfer of plasmid between the E. coli strains was determined.

Fermented milk consumption had little effect on the populations of R388 or pUB2380 transconjugants in the anexic mice but the milk fermented with both LAB resulted in a slight increase in E. coli PG1 (R388) transconjugants. Long-term consumption of milk fermented with both LAB strains resulted in inhibition of occurrence and maintenance of E. coli PG1 (pCE325) transconjugants in the anexic mice. L. bulgaricus-fermented milk resulted in a reversible reduction in the numbers of E. coli PG1 (pCE325) transconjugants in mouse faecal samples. When L. bulgaricus and S. thermophilus were grown in BHI broth, the transconjugant lowering capabilities observed with the LAB plus fermented milk were reversed. Administration of L. bulgaricus and the combination of both strains grown in BHI broth increased the population of E. coli PG1 (pCE325) transconjugants recovered from mouse faecal samples. Thus, the authors concluded that probiotic LAB have an effect on the formation and subsequent maintenance of transconjugants in the gastrointestinal tract of anexic mice, and that this effect was dependent on whether the LAB were administered in fermented milk or labouratory growth media, indicating that bacterial metabolic end products as well as viable LAB may be involved.

Clearly much work remains to be done to elucidate the role played by specific environmental parameters, such as diet, ratio of donor and recipient cells, activity of donor and recipient and antagonistic effects of the human gut microflora on plasmid transfer and persistence of GMOs which may be ingested in a viable form in fermented foods. Genetic engineering is now being employed by the food industry to improve the performance of microorganisms used in the production of fermented foods. Concern has been expressed over the exposure of consumers to genetically modified microorganisms which may be ingested in a viable form in future food products. Of chief concern is the risk of recombinant DNA transfer from ingested GMOs to members of the human gut microflora. A range of natural DNA transfer mechanisms are available to microorganisms in the environment. Chief among these are conjugation, transformation and transduction.

Apart from one case of transformation in the oral cavity, most studies examining DNA transfer in the human gastrointestinal microflora have been concerned with conjugation. In vitro and in vivo models of varying degrees of complexity have been employed to simulate the prevailing ecological conditions in the human gastrointestinal microflora. Such studies have demonstrated the transfer of plasmid DNA, both naturally occurring and genetically modified, in the digestive tract.

Some of the factors thought to affect conjugation between bacteria in the gut microflora have also been studied in vitro and in vivo, e.g. metabolic burden of plasmid maintenance, recombinant DNA stability, colonisation ability of GMOs and antagonistic activities of the microflora, presence or absence of antibiotics, cell concentrations of bacteria involved in conjugation and the effect of dietary supplements on DNA transfer events in the human gut microflora. In order to provide information of direct relevance to risk assessment of the use of genetically modified microorganisms in food, further scientific investigations employing in vivo and in vitro models.

EQUIVALENCE OF GENETICALLY MODIFIED CROPS

A very promising strategy to evaluate the equivalence of genetically modified crops is gene expression profiling. Conventional methods for the analysis of differential gene expression include Northern blotting, S1 nuclease protection, comparative expressed sequence tag (EST) sequencing, differential display and serial analysis of gene expression. More recently, Kok *et al.* have designed a method of detecting altered gene expression by means of mRNA fingerprinting or reverse transcription polymerase chain reaction (RT-PCR). Based on the original concept of Liang and Pardee specific subsets of the mRNA population of a tomato plant were amplified. This enables the expression of different genes in the GM crop and genes in the parental line to be compared and any empirical information about significant alterations to be established.

The method is still under development and requires further validation, but has been shown to be reproducible and able to detect important differences in gene expression. Some of the most significant remaining problems are that the method largely depends on individual skills based on experience and is time-consuming, even in its most simplified form.

CDNA MICROARRAYS

Over the past few years, the DNA microarray technology has emerged as a powerful high-throughput method for the analysis of gene expression. At present, microarrays for gene expression studies generally consist of cDNAs, which are physically deposited onto small glass surfaces. The major advantage of DNA microarray technology over conventional gene-profiling techniques is that it allows small-scale analysis of expression of a large number of genes in a sensitive and quantitative manner.

Furthermore, it allows comparison of gene expression profiles under a large number of different conditions. The cDNA microarray technology has already been successfully introduced in disciplines ranging from cell and developmental biology to drug development and pharmacogenomics. In the following a number of technical aspects of the technology as well as its potential value for the safety assessment of genetic modification of food plants.

The cDNA microarrays are produced by deposition onto a solid surface of purified PCR products corresponding to specific genes. Typically a few nanolitres of DNA solution (100-500 μg/ml) are spotted using a micro-dispensing robot. Essential information on robot construction and protocols can be found on the Internet. In general, with respect to the spotting devices, a higher density of spots is accomplished when using a passive dispenser (> 2500 DNA spots/cm2).

Thereto, glass slides covered with a positively charged layer (*e.g.* poly-l-lysine) or carrying reactive groups are most commonly used. To analyse the expression of genes in a plant cell or tissue sample, polyA+ RNA (typically 0.5-2.5 μg is required for each reaction) is purified and labelled by reverse transcription using an oligo-dT primer incorporating a fluorescently labelled nucleotide. The labelled cDNA pool is then hybridized to the microarray. The required hybridization conditions (*i.e.* sample concentration, stringency of hybridization) and parameters, such as detection level and the correlation between transcript concentration and hybridization have been determined for each experimental setting.

The intensity of a hybridization signal, which is a measure of the relative expression level of each gene, can be read by charge-coupled device (CCD) cameras or confocal/non-confocal laser scanners. RNAs extracted from two different samples can also be simultaneously analysed with a microarray by labelling each with a different fluorescent label using either Cy-3 or Cy-5. At RIKILT-DLO we are currently testing whether the analysis of differential gene expression using DNA microarrays is an informative strategy for the safety evaluation of GM plants.

To address this question, DNA libraries are constructed and enriched by a subtractive cloning approach for cDNAs preferentially expressed in either green or red tomato fruit. These subtracted cDNAs, as well as control cDNAs representing known tomato genes (*i.e.* sequence information derived from public sequence databases), have been spotted on arrays and are used for comparison of gene expression profiling of control and genetically modified tomatoes.

In the near future we will construct more defined arrays containing genes of toxicologically relevant pathways, such as those involved in the synthesis of natural plant toxins (*e.g.* the glycoalkaloids α-tomatine and α-solanine).

CHEMICAL FINGERPRINTING

The principles for establishing a chemical fingerprint were directed towards the detection of alterations in compositions in 1H-NMR spectra arriving from two populations, for example, non-modified counterpart(s) and an engineered one. In this way the unprocessed parts of the plant used for human consumption were fractionated and analysed, *i.e.* in tomato comparison on the fruit level.

Genetically modified tomato varieties, such as the antisense RNA exogalactanase fruit, have been studied using this innovative technology. The spread of measurements of the individual profiles was comparable for the GMO and its controls. By subtraction of the 1H-NMR spectra the differences of the means were calculated at 99% confidence intervals between the GM crop average (n = 8/batch per line) and its isogenic control average (n = 8/batch per line). The difference of the means was defined here as all differences of the means of the amplitudes of individual NMR signals obtained from spectra of various populations. The differences of the means were used to quantify how the transgenic crop plant differed from its parental line. The normalized 99% confidence intervals of the normalized means of fractions A-E after eightfold independent replicates demonstrated that differences in amplitudes exceeding at least 20% could normally be detected. This is an adequate sensitivity to allow statistical evaluation, and is in line with recommendations of the Nordic Council. They proposed that if the average value of a parameter differs by more than 20% an explanation should be sought.

With the purpose of establishing those differences that may arise from metabolic effects linked to genetic modification the default statistical analysis used a mixed effects randomized block model. Random block effects such as variations in the stage of cultivation, location, logistics or climate had a significant impact on the overall chemical fingerprints of cultivars. For example, chemical fingerprints varied much more between field sites and seasons than between replicates in one plot.

A tomato background mean (*i.e.* crop mean value) was established by constructing an extended range of non-modified tomato references for comparison with natural variability, and to assess the impact of external factors (*i.e.* variations in season, climate, etc.). The crop mean value was used to quantify how the differences of the means between the transgenic crop plant and its control differed from the natural diversity of commercial counterparts.

To differentiate between compositional changes either due to genetic modification, genetic variability or environmental variations the chemical fingerprints were studied in a hierarchical approach.

Hence, the fingerprints of the GMO are compared to those of:

- The isogenic control line grown side-by-side under identical conditions in one plot;
- The isogenic control line grown side-by-side under identical conditions at multiple sites;
- An extended range of commercial varieties of that crop;
- The influence of downstream processing.

It was recognized that 45% of all compounds of the control line harvested in different years varied considerably in concentration. Most differences were found in fractions A, C and D. A similar degree of impact on composition was observed in the case of processing temperatures (*e.g.* 100°C versus 70°C) on tomato juice. These findings were compatible with the protease activity surviving the 70°C treatment. But in fruit harvested from various individual plants, divided into two populations, approximately 95% of all constituents were practically identical in concentration. Also the determination of the differences of the means between transformant tEG to its isogenic control bred side-by-side in the same plot showed that 249 out of 3000 amplitudes (8%) varied in concentration. Assignments by NMR indicated that, for example, citric acid was present in fraction A of the control at 1.4 times the levels found in the transformant. However, in fraction A of the transformant glutamic acid and/or glutamine showed an increase of 1.3 times the levels found in the isogenic control.

On the other hand, one single aromatic compound, α-lycopene, appeared to be present in fraction E of transformant tEG at 2.5 times the concentrations found in the parental line. No alterations were observed in ß-carotene levels. These effects may be the result of the antisense downregulation of the tEG1A gene in tomato. In our view the natural diversity must be taken into consideration, however, when interpreting the biological relevance of any significant difference found in the differences of the means between the GMO and its parental line. The parental line was extended by entering non-isogenic commercial references (*i.e.* crop mean value).

Although limited in the range of lines, the comparison of the crop mean value to the differences of the means showed that the contribution and magnitude of differences (*e.g.* citric acid, glutamic acid/glutamine, α-lycopene) diluted out. It was recognized that in the agricultural practice the conditions of culturing, processing and storage of appropriate references will be mostly unknown and they may have a non-isogenic genotype. Notwithstanding these complicating random block effects we recommend that, for example, the chemical fingerprint should not be limited to the transgene and its isogenic parent bred under identical circumstances only, but should also include a sampling through multiple sites and a comparison with an extended range of controls of that crop.

Otherwise it would be a too limited approach towards the establishment of the claim of substantial equivalence and it would neglect the ranges of natural variations in that crop. In future, if many more lines have to be compared with each other this approach may benefit from multivariate statistical methods.

PROTEOME PROFILING

The introduction of Bt genes of bacterial origin may theoretically influence the nature of post-translation modification (*e.g.* glycosylation) in the transgenic tomato fruit. This is because as well as the impact of the protein moiety, for example, the type of glycosylation pattern may alter due to inactivation of glycosidases and glycosyltransferases in the host cell. In addition, glycosylation of the inserted Bt gene product in the plant might result in altered protein activity and stability.

The importance of structurally elucidating these oligosaccharides, especially the N-glycans, can be attributed to their widespread functions in the cell. These functions include correct folding, biological activity and stability of proteins as well as involvement in plant development. Moreover, N-glycans represent antigenic epitopes by themselves. Immunological as well as biochemical studies revealed the allergenic potency for the α1,3-fucosylation of the proximal N-acetylglucosamine residue. Additionally, also the β1,2-xylosylation of the β core mannose is suspected to be an allergenic epitope. Recently, there is increasing evidence that these N-linked oligosaccharides may have a substantial impact on the immunogenicity of glycoproteins, as N-glycans are suspected to play a certain role in the adverse food reactions of hypersensitized patients. At the Institute for Agrobiotechnology, Austria, this issue was tackled by developing a generic technology for isolating, purifying and characterizing N-linked oligosaccharides in plant tissues. Our low-temperature acetone powder method appeared to be most suitable for the isolation of (glyco-)proteins from plant tissue (*e.g.* tomato fruit). The resulting (glyco-)protein fraction was proteolytically digested and corresponding glycopeptides purified by cation-exchange chromatography.

The oligosaccharides (N-glycans including α1,3-fucosylated oligosaccharides) were released from the peptide backbones by N-glycosidase A (PNGase A) and fluorescently labelled with 2-aminopyridine. Subsequently, pyridylaminated glycans were separated and analysed structurally by repeated two-dimensional high-performance liquid chromatography (2D HPLC) using a size-fractionation and a reverse-phase column in combination with an exoglycosidase treatment.

MALDI-TOF (matrix-assisted laser desorption ionization time-of-light) mass spectrometry was applied for further confirmation of the observed structures. Sixteen different N-glycosidic structures could be detected in tomato fruit. The two most abundant glycans in all the tomato samples showed identical properties to those of the major N-linked oligosaccharides of horseradish peroxidase (MMXF: Man(3)Xyl-FucGlcNAc(2)) and pineapple stem bromelain (MOXF: Man(2)Xyl-FucGlcNAc(2)), respectively, and accounted for about 65-78% of the total glycan content.

Oligomannosidic glycans occurred in only small quantities (3-9%). The majority of the N-glycans found was β 1,2-xylosylated and carried an α1,3-fucose residue linked to terminal N-acetylglucosamine. A structural element was identified and shown to be an IgE-reactive determinant, which contributes to cross-reactivities among non-related glycoproteins, and is present in a wide variety of plants extracts.

It is proposed that this structural element might also be of importance in some cases of food and pollen allergy. The carbohydrate profiling of, for example, populations of green and red-ripe fruit of non-modified and modified Bt tomato plants (expressing the cryIA5 gene) revealed no differences in the nature of N-glycans, nor were there significant differences in the relative amounts of glycan. Interestingly, the relative amounts found were quite similar across seasons and comparable to the commercial variety Notoro used as an extended control. As far as the N-linked oligosaccharides were involved it was not possible to detect significant alterations in the α1,3-fucose content of Bt tomatoes due to genetic modification.

PROFILING OF THE METABOLOME

Among other factors, identity of a complex GM crop plant with its traditionally bred parent entails metabolism and the level of (un-)desirable substances contained therein. Therefore, we would like to build up an overall picture of components associated with different metabolic capabilities in the (non-)modified plant tissue. A multi-compositional analysis of the so-called metabolome may overcome the test limitations of, for example, single critical nutrient or key toxicant analyses and problems related to their limits of natural diversity.

An attractive alternative is the utility of high-resolution proton (1H) nuclear magnetic resonance (NMR) spectroscopy (1H-NMR) in combination with separation techniques (liquid chromatography (LC)). Recently, we have shown that the approach of chemical fingerprinting using off-line LC-NMR can obtain information on possible changes in complex plant matrices due to environmental effects as long as GLP-like conditions for sample handling, data acquisition and automation of data handling are maintained. In principle, it will be possible to screen all the different low molecular weight components (MW <10 kDa) present in the metabolome of the plant tissue.

TOXICOLOGICAL PROFILING

The significance of detected compositional alterations may need to be further explored by in vitro toxicity assays in order to screen for potential adverse effects and for mechanistic studies. It may even be that the latter approach leads to a redesign, refinement or even replacement of (sub-)chronic rodent feeding trials. In vitro systems derived from organs and tissues from animals and humans, and various types of cultured recombinant cell lines, are successfully used for screening for the toxic potential of single compounds.

However, their ability to screen for the potential toxicity of whole foods or extracts thereof has been insufficiently explored up till now, and has been examined by us within the framework of a tiered safety evaluation of GM plants. The validity of using in vitro test assays for comparative testing of whole food products, food ingredients and extracts thereof depends on the sensitivity and selectivity of the toxic response induced upon exposure. However, since these models are not yet validated, they should be considered as early warning or alert-systems to identify potential changes in toxicity of the engineered product only.

ASSESSING TRANSCRIPTION FACTORS

There is a demand for in vitro models that not only determine the (geno-)toxic potency of plant components or complex mixtures, but also provide information on the mechanism by which that compound exerts its toxic effect. As an in vitro profiling system the eukaryotic stress gene assay (*i.e.* CAT-Tox(L)) was assessed. This utilizes human HepG2 cells stably transfected with chloramphenicol acetyltransferase (CAT) reporter constructs. The aim of the CAT-Tox(L) assay is to collect and quantify molecular responses to cellular stress and toxicity that have a transcriptional component in their regulation. The assay employs an ELISA method to measure the transcriptional activity of the promoter or response element constructs. The advantage is that many of the stress responses occur before any measurable cytotoxicity, thus allowing monitoring of stress pathways at sublethal levels.

Since 14 unique transcription responses can be measured simultaneously, the CAT-Tox(L) provides comprehensive cellular stress profiles. The assay generates specific stress gene induction profiles for a variety of stressful and toxic plant components such as genistein, diadzein and sinigrin.

Aqueous extracts of red-ripe, non-and antisense RNA exogalactanase tomato fruit dosed up to 0.05 g per assay did not cause any cytotoxicity to the HepG2 transfectants nor were there molecular responses to cellular stress or toxicity. Extracts of non-modified and modified green tomato fruit showed cytotoxicity and induced the construct containing the xenobiotic response element (XRE).

This gene-induction profile was not comparable to that of, for example, the reference molecules dioxin, 3-methylcholanthrene, benzo[a]pyrene or β - napthoflavone. These polycyclic aromatic hydrocarbons were characterized with a co-induction of cytochrome P450 1A1 and glutathione-S-transferase Ya subunit- CAT fusion constructs. There was substantial evidence that the systemic fungicides fenarimol and bupirimate did not induce any of the genes in this assay. There was no promotor in the assay that could be linked to universal cellular stress.

A-tomatine caused severe cytotoxicity at high concentrations and yet did not induce any of the genes (*i.e.* causing membrane damage). However, it was observed that α-tomatine diluted in a matrix of red-ripe tomato tissue (final extract concentration 5% w/v) elicited an increased cytotoxicity concomitant with a dose-dependent response profile involving XHF (collagenase involved in inflammatory reactions sensitive to mitogens and cytokine IL-1). So far, the approach of the CAT-Tox(L) model shows potential but there is a need for validation based on molecular mechanisms of plant constituents.

Although the induction of these stress genes represents cellular perturbations, the mechanistic validations need to be extended before any conclusions can be drawn on the value of these findings. Certainly, the development of a database to explore relationships between a wide variety of compounds stressful to plants and meaningful stress gene profiles may serve as molecular fingerprints unique for a particular component.

GENOTOXICITY ENDPOINTS

The mucosa of the gastrointestinal tract is clearly a primary, potential site of action for the effects of novel foods. Therefore effort has been concentrated on the development of a battery of intestinal cell systems (*e.g.* IEC-6, IEC-18, Caco-2 and INT407 cell lines) with different endpoints for initial cyto- and genotoxicity testing. The choice of a suitable method of sample preparation was considered of utmost importance in assaying vegetables.

Freeze-dried tomato fruit was extracted (10% suspensions, w/v) in water, 0.9% saline or chloroform:methanol (2:1, v/v) immediately before incorporation into the culture medium. It was encouraging that aqueous and chloroform/methanol extracts of, for example, red-ripe tomatoes up to a maximum of 10% suspensions (w/v), had little or no general toxicity once the pH (> 6.0) and osmotic pressure had been taken into account.

No toxicity to the intestinal cells Caco-2 and IEC-18 under short-term conditions was detected in red-ripe, non- and antisense RNA exogalactanase tomato using alpha-glutathione S-transferase (α-GST) lactate dehydroyrenase (LDH) leakage, reduction of 3-(4,5-dimethylthiazol-2-yl)-2,5 diphenyltetrazolium bromide (MTT conversion), neutral red (NR) uptake and total cellular protein as in vitro endpoints.

On the other hand, green tomato fruit displayed a pronounced toxic and cell-damaging effect. No statistical significant differences were found between transformants and isogenic control lines in terms of their activity in the assays. Treatment of tomato tissue with pepsin (ratio pepsin to tomato protein 1:250, w/w) did not result in an increase of or loss of inherent cytotoxicity. Supplementary investigations with naturally occurring plant components, for example, quercetin, β -carotene, α-lycopene, genistein showed that only steroid glycoalkaloids, such as α-tomatine, exhibited a rapid toxic effect, whereas α-tomatidine was ineffective. It was noted that the in vitro models did show general matrix-related toxic effects. For example, α-tomatine diluted in a matrix of extracted red-ripe tomato tissue (final concentration of 5-10% w/v) showed an increase of dose-response-dependent cytotoxicity in human Caco-2 cells (*e.g.* IC50 7.9 ± 2.1 mg/L versus IC50 33.9 ± 5.3 mg/L of α-tomatine only). Using the COMET assay, suspensions of freeze-dried tomato fruit up to a maximum of 10% (w/v) did not exhibit DNA-damaging effects in a range of gastrointestinal cells derived from rats (IEC-6 and IEC-18) and humans (Caco-2, INT407).

Genetic modification did not result in increased genotoxicity, but fruit extracts were able to suppress the DNA-damaging effects of the known genotoxins H_2 O_2 and N-methyl-N_2 -nitro-N-nitrosoguanidine (MNNG). Although these endpoints are important toxicological measures they do not indicate the molecular mechanisms of toxic damage. So far, pronounced adverse effects could only be observed with glycoalkaloids. The screening for subtle differences in toxic responses of other plant constituents appeared to be limited. However, therefore, the use of critical endpoints for such studies other than cytotoxicity should be clearly defined.

BIOTECHNOLOGY IN THE PRODUCTION OF FOOD INGREDIENTS

As described in the Introduction, flavouring agents, organic acids, food additives and amino acids are all metabolites of microorganisms during fermentation processes. Microbial fermentation processes are therefore commercially exploited for production of these food ingredients. Metabolic engineering, a new approach involving the targeted and purposeful manipulation of the metabolic pathways of an organism, is being widely researched to improve the quality and yields of these food ingredients. It typically involves alteration of cellular activities by the manipulation of the enzymatic, transport and regulatory functions of the cell using recombinant DNA and other genetic techniques. Understanding the metabolic pathways associated with these fermentation processes, and the ability to redirect metabolic pathways, can increase production of these metabolites and lead to production of novel metabolites and a diversified product base.

BIOTECHNOLOGY IN DIAGNOSTICS FOR FOOD TESTING

Many of the classical food microbiological methods used in the past were culture-based, with microorganisms grown on agar plates and detected through biochemical identification. These methods are often tedious, labour-intensive and slow. Genetic based diagnostic and identification systems can greatly enhance the specificity, sensitivity and speed of microbial testing. Molecular typing methodologies, commonly involving the polymerase chain reaction (PCR), ribotyping (a method to determine homologies and differences between bacteria at the species or sub-species (strain) level, using restriction fragment length polymorphism (RFLP) analysis of ribosomal ribonucleic acids (rRNA) genes) and pulsed-field gel electrophoresis (PFGE, a method of separating large DNA molecules that can be used for typing microbial strains), can be used to characterise and monitor the presence of spoilage flora (microbes causing food to become unfit for eating), normal flora and microflora in foods. Random amplified polymorphic DNA (RAPD) or amplified fragment length polymorphism (AFLP) molecular marker systems can also be used for the comparison of genetic differences between species, subspecies and strains, depending on the reaction conditions used. The use of combinations of these technologies and other genetic tests allows the characterisation and identification of organisms at the genus, species, sub-species and even strain levels, thereby making it possible to pinpoint sources of food contamination, to trace microorganisms throughout the food chain or to identify the causal agents of foodborne illnesses. Monoclonal and polyclonal antibodies can also be used for diagnostics, *e.g.* in enzyme-linked immunosorbent assay (ELISA) kits.

Microarrays are biosensors which consist of large numbers of parallel hybrid receptors (DNA, proteins, oligonucleotides). Microarrays are also referred to as biochip, DNA chip, DNA microarray or gene arrays and offer unprecedented opportunities and approaches to diagnostic and detection methods. They can be used for the detection of pathogens, pesticides and toxins and offer considerable potential for facilitating process control, the control of fermentation processes and monitoring the quality and safety of raw materials.

FACTORS

This conference deals with the application of biotechnology to food processing in developing countries. Biotechnological research as applied to bioprocessing in the majority of developing countries, targets development and improvement of traditional fermentation processes. In this section, we consider some areas specifically relevant to developing countries and list some key issues that should be considered by participants in this conference.

SOCIO-ECONOMIC AND CULTURAL FACTORS

Traditional fermentation processes employed in most developing countries are low input, appropriate food processing technologies with minimal investment requirements. They make use of locally produced raw materials and are an integral part of village life. These processes are, however, often uncontrolled, unhygienic and inefficient and generally result in products of variable quality and short shelf lives. Fermented foods, nevertheless, find wide consumer acceptance in developing countries and contribute substantially to food security and nutrition.

How will applications of biotechnology to fermented foods impact on these socio-economic and cultural factors?

INFRASTRUCTURAL AND LOGISTICAL FACTORS

Physical infrastructural requirements for the manufacture, distribution and storage (*e.g.* by refrigeration) of microbial cultures or enzymes on a continuous basis is generally available in urban areas of many developing countries. However, this is not the case in most rural areas of developing countries.

Should research be oriented to ensure that individuals at all levels can benefit from applications of biotechnology in food fermentation processes, *i.e.* should logistical arrangements for starter culture development be integrated into biotechnological research targeting improvement of traditional fermentations? What is required for the level of fermentation technologies and process controls to be upgraded in order to increase efficiency, yields and the quality and safety of fermented foods in developing countries?

NUTRITION AND FOOD SAFETY

Fermentation processes enhance the nutritional value of foods through the biosynthesis of vitamins, essential amino acids and proteins, through improving protein and fibre digestibility; enhancing micronutrient bioavailability and degrading antinutritional factors. Many bacteria in fermented foods also exhibit functional properties (probiotics). The safety of fermented food products is enhanced through reduction of toxic compounds, such as mycotoxins and cyanogenic glucosides, and production of antimicrobial factors, such as bacteriocins, carbon dioxide, hydrogen peroxide and ethanol, which facilitate inhibition or elimination of food-borne pathogens.

Are the nutritional characteristics (and safety aspects) of fermented foods adequately documented and appreciated in developing countries? Is there a need for consumer education about the benefits of fermented foods?

INTELLECTUAL PROPERTY RIGHTS (IPRS)

The processes used in the more advanced areas of agricultural biotechnology tend to be covered by IPRs and these rights tend to be owned by parties in developed countries. This applies also to biotechnology processes used in food processing. On the other hand, many of the traditional fermentation processes applied in developing countries are based on traditional knowledge.

In addition to biotechnology processes, microbial strains may also be the object of IPRs. For example, an era of massive private investment in biotechnology was initiated when the United States Supreme Court ruled in 1980 (in the Diamond versus Chakrabarty case) that a live GM bacterium (of the genus *Pseudomonas*, modified to degrade components of crude oil) could be patented. Many of the microorganisms associated with traditional fermentation processes in developing countries are unique. Issues of ownership will become increasingly important as bacterial strains are characterised and starter cultures are developed in developing countries.

How should food scientists, researchers, industry and governments in developing countries approach these issues? - A considerable volume of research into the development and improvement of fermentation processes is currently taking place worldwide. Are the research results from developing countries adequately documented? Who owns this information? Are cell banks being developed to protect microbial strains characterised in developing countries?

COMMERCIAL OPPORTUNITIES

Biotechnological innovations have greatly assisted in industrialising production of certain indigenous fermented foods. Indonesian tempe and Oriental soy sauce are well known examples of indigenous fermented foods that have been industrialised and marketed globally. The results of biotechnology research will lead to fermented foods of improved quality, safety and consistency.

AN ENERGY TRANSITION FOR SARD AND FOOD SECURITY

During the last five years a new context for action has emerged in the international community, which calls for an energy transition in both developing and industrialized countries. For the OECD countries this transition takes form in the context of high consumption patterns that involve net resource transfers from South to the North, consume excessive resources and/or generate pollution.

In the African countries, a transition to sustainable energy systems is needed to accelerate the growth of basic food production, harvesting and processing. However, breaking the current energy bottleneck must also be sustainable (*viz.* environmentally sound, socially acceptable and economically viable). Such a transition involves a commitment to long-term developmental goals and requires innovative policy and technological solutions.

For Africa, an energy transition would be characterized by a move from the present levels of subsistence energy usage based on human labour and fuelwood resources, to a situation where household, services and farming activities use a range of sustainable and diversified energy sources. Obvious benefits are greater resilience in the production system, higher productivity, improved efficiency and higher incomes to farmers.

Environmental degradation, driven primarily by poverty, would be minimized. The investment required to make such a transition would not be significantly different from that required for conventional approaches.

However, the process of identifying needs and promoting investment in a range of technological options would be considerably different. The new energy situation would offer opportunities to reinvigorate the situation in many African countries which continue to cope with insufficient rural energy supplies.

Among the problems are the following:

- Price policies rarely reflect the energy needs of rural populations;
- Energy plans and agricultural programmes are not linked;
- Energy requirements for agro-activities are seldom quantified;
- Energy policies and plans in most countries do not focus on the agricultural and rural sectors, except occasionally, on an aggregate basis;

Unfortunately, the energy transition in rural areas will not occur under a "business-as-usual" situation. A concerted effort is needed on the part of the many actors influencing energy supply and demand patterns. One challenge is to reduce the barriers facing rural energy development which arise from a lack of policy and programme coordination between the rural and agricultural sectors, and the energy sector institutions.

The rural sector continues to remain outside the energy assessment and planning efforts which are normal practice for industry, commerce, transport. This is due, in part, to the small impact rural energy has on the national energy balance. Because of their meagre energy consumption and poor data, especially in the African context, the important role agriculture plays as a source of food and fibre, for foreign exchange and the relatively large percentage of GDP that is derived from the agricultural and rural sector, is ignored. The dispersed and often non-monetized nature of rural energy also contributes to its neglect in planning and investment. Energy authorities rarely have an institutional or operational presence in rural areas and only a few agriculture and rural development programmes deal explicitly with rural energy requirements. This is due, in part, to lack of technical capability. However, a change in mindset is also needed among policy makers to recognize the potential economic and social gains to be realised from increasing energy supply in rural areas. These gains will translate into improved use and management of land resources by allowing more efficient use of resources and less degrading land-use practices such as fuelwood use.

National agricultural and rural development authorities, normally without any mandate regarding energy matters, are often incapable of negotiating their energy requirements with electricity utility companies and energy authorities.

Thus, a "vacuum" of responsibility and lack of guidance for energy interventions in rural areas seems to exist in most countries. No institution is actually "in charge" of energy for development of the rural and agricultural sector. This leads to low allocation of resources and investment for rural development and agricultural activities vis-à-vis other sectors of the economy. Since no single institution, either governmental, local or private could alone cope with all issues involved, a political interest, coupled with effective inter-institutional cooperation and collaboration is required.

Promoting food security by raising agricultural productivity and sustainable production systems will inevitably involve increases in energy inputs for water supply and management, plant nutrients, and agro-processing, to provide community lighting and drinking water, for cultivation. Consistent with the SARD framework mentioned earlier in this report is the need to shift the emphasis from single issue solutions to more integrated, sustainable approaches to development.

For example, pesticides alone are not sufficient or economically cost-effective in controlling most pest problems. Strategies now exist for many crops which involve understanding the pest life cycle, economic damage thresholds and the effects of cultivation practices which can greatly reduce or even eliminate the need for regular pesticide applications. IPM is an effective way of reducing production costs and avoiding the associated risk of pollution and contamination.

Similar evidence of the benefits from integrated approaches exists for mineral fertilizers. Integrated plant nutrition strategies that use organic materials, leguminous crop rotation, and cultivation practices to maintain the optimal balance of soil structure and plant nutrients for agriculture are more beneficial economically to farmers than solely relying on mineral fertilizers. The availability of adequate water resources for agriculture is essential for increased production. From 1960-1980, water resources in Africa have declined from 16.5 million cubic metres per capita to 9.4. By the year 2000 this is expected to 5.1.

It is projected that by the end of the 1990s six out of seven east African countries and all five north African countries bordering the Mediterranean sea (Algeria, Egypt, Libya, Morocco and Tunisia) will face acute water shortages; with six of the countries having less than 1 million cubic metres per person. However, efficient use of this resource in Africa does not imply large scale, energy-intensive irrigation schemes. Small pumps have had an important beneficial effect on irrigation in some African countries of vegetable and even rice production. Where surface water is available this technology represents a well distributed and energy efficient option.

Previous experience in Africa has shown that the way in which the water resource is made available, both its price and mode of delivery, will determine whether the resource is used sustainably. Thus irrigation schemes should follow the principles established by the International Action Programme on Water and Sustainable Agricultural Development which takes into account the planning, development and management of water resources in an integrated manner. Also important is the potential for biomass energy conversion technologies. Residues from wood and agro-industries, purposely grown biomass and municipal solid wastes may play a major role in many African countries. The economic and social assessment of these options is needed to avoid disrupting employment and resource use. Local and global environmental benefits of biomass energy conversion must also be considered.

As a result of UNCED, efforts are underway to develop national Agenda 21's. These provide valuable entry point for countries to fully integrate rural energy requirements and potentials into their energy and SARD strategies. They involve multidisciplinary and convergent actions at both national and local levels, which draw together energy, agricultural and environmental knowledge, experience and policies. National and district rural energy strategies are needed to provide a common framework and plan to direct investment and pull together the efforts of various ministries such as Agriculture, Energy, Forestry, Planning and Agrarian Reform. NGOs, local community groups, and the private sector have an important role to play in such initiatives.

5

Applications of Plant Genetic Engineering

"PHARMING" AND "PLANTIBODIES"

An increasingly viable option is the production of highly valuable enzymes by plants and animals. In addition to production of human proteins in these organisms, other valuable proteins that are currently produced by microorganisms could very well be produced by higher organisms instead.

Animal and plant "bioreactors" in some respects may be superior to recombinant bacterial systems, because eukaryotes glycosylate proteins. Whereas the glycosylation pattern may be species-specific, appropriate glycosylation is often required for protein function. Production through these organismal systems may also be cheaper than cell fermentation techniques.

Two examples of production of human proteins in plants include the production of human serum albumin in transgenic tobacco and potato, and production of human insulin by tobacco. In both cases, the produced protein appears to be fully effective in humans. Unfortunately, however, one cannot raise his/her insulin level by eating transgenic tobacco leaves, as the protein in most cases will be broken down to amino acids before it reaches the blood stream.

Therefore, in these cases one cannot escape the practice of protein isolation and purification before transgenic leaves are converted into drugs. Also antibodies are being produced in plants. Initially, the antibody's light and heavy chains were produced in different plants. But a subsequent cross of these two varieties resulted in progeny carrying assembled and functional antibodies.

REVERSIBLE MALE STERILITY IN PLANTS

As has been indicated earlier, heterozygous individuals often are healthier and stronger than homozygous ones. The only way to guarantee heterozygoiscity in plants is to make sure self-pollination cannot occur. For most crop plants it was very tedious or practically impossible to exclude selfing. To exclude self-pollination, it would be good to introduce male sterility in plants: progeny from such plants are then expected to be 100percent heterozygous (assuming they were pollinated with pollen from an unrelated variety). To introduce male sterility, a promoter was identified that was turned on exclusively in tapetum cells (a tissue around the pollen sac that is essential for pollen production). This promoter then was linked up to a gene coding for a bacterial ribonuclease (named barnase). This ribonuclease selectively chops up ribonucleic acids. The promoter/ribonuclease construct was then introduced into plants (canola, tobacco, you name it).

Because the promoter allows expression only in tapetum cells, the gene construct disrupts only development of the tapetal tissue and its end product, pollen. Plants transformed with this construct were male-sterile but otherwise normal. Although male-sterile plants are valuable for hybrid seed production, they have limited value when it comes to crop production. Fertility must be restored to crops such as wheat, rice, and tomato, in which the seed or fruit is the harvested product. Fortunately, the ribonuclease is inhibited very much by a simple protein, named barstar. One can thus cross the male-sterile plant with a male-fertile variety in which the gene for barstar has been introduced, and the result is progeny with viable pollen and restored fertility. A closely related approach has been criticized as "terminator technology" as it is seen by its critics as a way for companies to protect and enforce their patents.

ANTISENSE RNA

Antisense RNA refers to nucleotide strands that are produced in a cell and that are complementary to a particular mRNA. Antisense RNA can be produced, by inverting the coding region of a gene with respect to its promoter. The antisense RNA can hybridize with its corresponding mRNA, making it double-stranded. The double-stranded mRNA no longer can be recognized by the protein-synthesizing machinery (the ribosomes), and thus expression of this mRNA is suppressed.

Also, in many systems double-stranded mRNA is very unstable and is broken down quickly. Thus, one can inactivate specific genes while not interfering with others. Antisense approaches already are used to protect plants from damage by plant viruses. Reversal of a gene from bean yellow mosaic virus (BYMV), and putting it into tobacco under a reasonably strong promoter, has led to a tobacco variety that is quite resistant to BYMV.

A similar approach is used to transfer viral resistance to other plants. This finding is of significance, in that currently no effective, environmentally friendly methods exist to control many plant viruses. Very related to this approach is the RNAi (RNA interference) approach. This is very useful for both agriculture and medicine, and the first examples of practical applications of this RNAi technology are appearing. As with any new technology, the initial pilot projects are sort of pedantic. However, more exciting application possibilities abound. Obviously, RNAi technology provides an excellent approach for reverse genetics in eukaryotes. With this method one can turn off genes.

Perhaps indicative of the large potential and relative ease of genetic engineering, developing countries (particularly China) are progressing rapidly in development and application of genetically engineered crops. Some have gone into commercial production well ahead of similar crops in the US. In China, tomatoes that have been engineered for improved virus resistance have been on the market since late 1992. There are two main reasons for the more rapid commercialization of bioengineered crops in the developing world:

- Less tight governmental approval mechanisms
- Hungrier populations.

While in developed countries the main value of biotechnological applications may be to reduce production costs, in the developing world a main factor is the production of more food. Indeed, genetic engineering applications seem to be pretty successful to cut down on pathogen-induced losses. Genetic modification of papaya plants (expression of the ringspot virus coat protein in the plant) protects very well against the very destructive ringspot virus.

CHARACTERISTICS OF PLANT GENETIC ENGINEERING

Plant cells have a variety of characteristics that distinguish them from animal cells. These characteristics include the presence of a large central vacuole and a cell wall, and the absence of entioles, which play a role in mitosis, meiosis, and cell division.

Along with these physical differences, another factor distinguishes plant cells from animal cells, which is of great significance to the scientist interested in biotechnology: Many varieties of full-grown adult plants can regenerate from single, modified plant cells called protoplasts - plant cells whose cell walls have been removed by enzymatic digestion.

More specifically, when some species of plant cells are subjected to the removal of the cell wall by enzymatic treatment, they respond by synthesizing a new cell wall and eventually undergoing a series of cell divisions and developmental processes that result in the formation of a new adult plant. That adult plant can be said to have been cloned from a single cell of a parent plant.

Crop genetic engineering includes:

- DNA isolation
- Gene cloning
- Gene design
- Transformation,
- Plant breeding.

Plants that can be cloned with relative ease include carrots, tomatoes, potatoes, petunias, and cabbage, to name only a few. The capability to grow a whole plant from a single cell means that researchers can engage in the genetic manipulation of the cell, let the cell develop into a completely mature plant, and examine the whole spectrum of physical and growth effects of the genetic manipulation within a relatively short period of time. Such a process is far more straightforward than the parallel process in animal cells, which cannot be cloned into full-grown adults. Therefore, the results of any genetic manipulation are usually easier to examine in plants than in animals.

DNA CLONING

DNA cloning a method to manipulate DNA in vitro and in vivo or is a logical extension of the ability to manipulate DNA molecules with restriction endonucleases and ligases. Imagine that an plant gene has been obtained as a single restriction fragment after digestion of a larger molecule with the restriction enzyme *Bam*HI, which leaves 5-GATC-3 sticky ends. Imagine also that a small *E. coli* plasmid has been purified and treated with *Bam*HI, which cuts the plasmid in a single position. The circular plasmid has therefore been converted into a linear molecule, again with 5-GATC-3 sticky ends. Mix the two DNA molecules together and add DNA ligase. Various recombinant ligation products will be obtained, one of which comprises the circularized plasmid with the animal gene inserted into the position originally taken by the *Bam*HI restriction site.

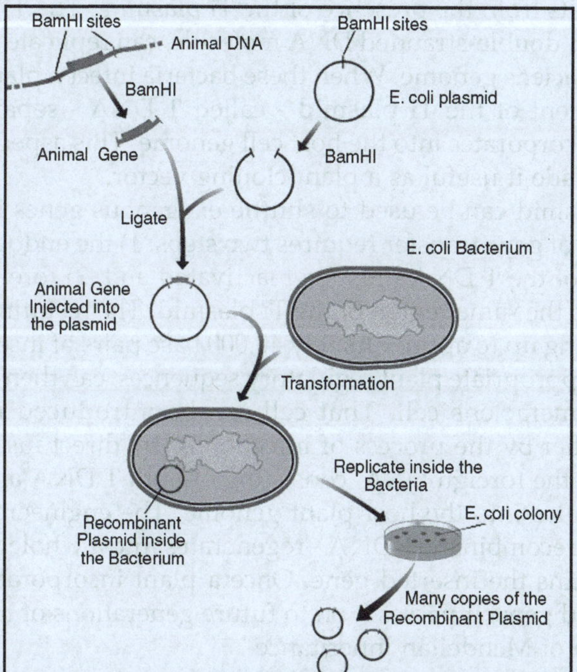

Fig. An Outline of Gene Cloning.

If the recombinant plasmid is now re-introduced into *E. coli*, and the inserted gene has not disrupted its replicative ability, then the plasmid plus inserted gene will be replicated and copies passed to the daughter bacteria after cell division.

More rounds of plasmid replication and cell division will result in a colony of recombinant *E. coli* bacteria, each bacterium containing multiple copies of the animal gene. This series of events, constitutes the process called DNA or gene cloning.

Vector that Works with Plant Cells

Not all aspects of the genetic manipulation of plant cells are readily accomplished. Not only do plants usually have a great deal of chromosomal material and grow relatively slowly as compared with single cells grown in the laboratory, but few cloning vectors can successfully function in plant cells. While researchers working with animal cells can choose among a wide variety of cloning vectors to find just the right one, plant cell researchers are currently limited to just a few basic types of vectors.

Perhaps the most commonly used plant-cloning vector is the "Ti" plasmid, or tumor-inducing plasmid. This plasmid is found in cells of the bacterium known as Agrobacterium tumefaciens, which normally lives in soil. The bacterium has the ability to infect plants and cause a crown gall, or tumorous lump, to form at the site of infection. The tumor-inducing capacity of this bacterium results from the presence of the Ti plasmid. The Ti plasmid itself, a large, circular, double-stranded DNA molecule, can replicate independently of the A. tumefaciens genome. When these bacteria infect a plant cell, a 30,000 base-pair segment of the Ti plasmid - called T DNA - separates from the plasmid and incorporates into the host cell genome. This aspect of Ti plasmid function has made it useful as a plant cloning vector.

The Ti plasmid can be used to shuttle exogenous genes into host plant cells. This type of gene transfer requires two steps: 1) the endogenous, tumor-causing genes of the T DNA must be inactivated and, 2) foreign genes must be inserted into the same region of the Ti plasmid. The resulting recombinant plasmid, carrying up to approximately 40,000 base pairs of inserted DNA and including the appropriate plant regulatory sequences, can then be placed back into the A. tumefaciens cell. That cell can be introduced into plant cell protoplasts either by the process of infection or by direct insertion. Once in the protoplast, the foreign DNA, consisting of both T DNA and the inserted gene, incorporates into the host plant genome. The engineered protoplast - containing the recombinant T DNA - regenerates into a whole plant, each cell of which contains the inserted gene. Once a plant incorporates the T DNA with its inserted gene, it passes it on to future generations of the plant with a normal pattern of Mendelian inheritance.

One of the earliest experiments that involved the transport of a foreign gene by the Ti plasmid involved the insertion of a gene isolated from a bean plant into a host tobacco plant. Although this experiment served no commercially useful purpose, it successfully established the ability of the Ti plasmid to carry genes into plant host cells, where they could be incorporated and expressed.

A. Limition of A. Tumifecian

The fact that only certain types of plants were naturally susceptible to infection with the host bacterial organism initially limited the usefulness of the Ti plasmid as a cloning vector. In nature, A. tumefaciens infects only dicotyledons or "dicots" - plants with two embryonic leaves.

Dicotyledenous plants, divided into approximately 170,000 different species, include such plants as roses, apples, soybeans, potatoes, pears, and tobacco. Unfortunately, many important crop plants, including corn, rice, and wheat, are monocotyledons - plants with only one embryonic leaf - and thus could not be easily transfected using this bacterium.

Overcoming the Limition

Research efforts in the past few years have reduced the limitations of A. tumefaciens. Scientists discovered that by using the processes of microinjection, electroporation, and particle bombardment, naked DNA molecules can be introduced into plant cell types that are not susceptible to A. tumefaciens transfection. Microinjection involves the direct injection of material into a host cell using a finely drawn micropipette needle.

Electroporation uses brief pulses of high voltage electricity to induce the formation of transient pores in the membrane of the host cell. Such pores appear to act as passageways through which the naked DNA can enter the host cell. Particle bombardment actually shoots DNA-coated microscopic pellets through a plant cell wall. These developments, important in the commercial application of plant genetic engineering, render the valuable food crops of corn, rice, and wheat susceptible to a variety of manipulations by the techniques of recombinant DNA and biotechnology.

PLANT TRANSFORMATION

Genetic engineering of plants is much easier than that of animals. There are several reasons for this:

- There is a natural transformation system for plants (the bacterium *Agrobacterium tumefaciens*),
- Plant tissue can redifferentiate (a transformed piece of leaf may be regenerated to a whole plant), and
- Plant transformation and regeneration are relatively easy for a variety of plants.

The soil bacterium *Agrobacterium tumefaciens* ("tumefaciens" meaning tumor-making) can infect wounded plant tissue, transferring a large plasmid, the Ti plasmid, to the plant cell. Part of the Ti (tumor-inducing) plasmid apparently randomly integrates into the chromosome of the plant.

The integrated part of the plasmid contains genes for the synthesis of (1) food for the bacterium, and (2) plant hormones. Genes from the Ti plasmid that are integrated in the plant chromosome are expressed at high levels in the plant. Overproduction of the plant hormones leads to continuous growth of the transformed cells, causing plant tumors. Rapid, cancerous growth of the transformed plant tissue obviously is advantageous to the bacterium: more food gets produced.

The Ti plasmid has been genetically modified ("disarmed") by deleting the genes involved in the production of bacterial food and of plant hormones, and inserting a gene that can be used as a selectable marker. Selectable marker genes generally are coding for proteins involved in breakdown of antibiotics, such as kanamycin. Any gene of interest can be inserted into the Ti plasmid as well. In principle, one can thus transform any plant tissue, and select transformants by screening for antibiotic resistance. However, unfortunately, there are some complications: (1) it has proven difficult to transform some monocots (grasses, etc.) by *Agrobacterium*, and (2) regeneration of plants from tissue culture or leaf discs is not always possible.

A number of genetically engineered plant varieties have been developed. Traits that have been introduced by transformation include herbicide resistance, increased virus tolerance, or decreased sensitivity to insect or pathogen attack. Traditionally, most of such genetically engineered plants were tobacco, petunia, or similar species with a relatively limited agricultural application. However, during the past decade it now has become possible to transform major staples such as corn and rice and to regenerate them to a fertile plant. Increasingly, the transformation procedures used do not depend on *Agrobacterium tumefaciens*.

Instead, DNA can be delivered into the cells by small, μm-sized tungsten or gold bullets coated with the DNA. The bullets are fired from a device that works similar to a shotgun. The modernized device uses a sudden change in pressure of He gas to propel the particles, but the principle of "shooting" the DNA into the cell remains the same. This DNA-delivery device is nicknamed "gene gun", and has been shown to work for DNA delivery into chloroplasts as well.

Over the last several years, use of the "gene gun" has become a very common method to transform plants, and has been shown to be applicable to virtually all species investigated. For example, transformation of rice by this method is now routine. This is a very important development as rice is the most important crop in the world in terms of the number of people critically dependent on it for a major part of their diet.

Another method to get foreign genes into cereals is by electroporation: a jolt of electricity is used to puncture self-repairing holes in protoplasts (*i.e.*, the cell without the cell wall), and DNA can get in through these holes. However, it is often very difficult to regenerate fertile plants from protoplasts of cereals. Nonetheless, significant advances in overcoming these practical difficulties have been made over the years. Now even transgenic trees have been created: for example, the gene for a coat protein of the plum pox virus has been introduced into apricot. The plum pox virus leads to the feared Sharka disease, for which there is no cure. The resulting transgenic tree shows a markedly decreased sensitivity to this virus. The reason why continuous exposure of the tree to the viral coat protein leads to tolerance against viral infection is not yet understood, however. Thus, now there are a number of different techniques to introduce foreign genes into plants. Essentially all major crop plants can be (and have been or are being) genetically engineered, the procedures are now routine and the frequency of success is very high. Even though genetically engineered crops are more costly than the usual ones, US farmers have rather readily accepted them provided that tangible benefits can be demonstrated.

However, it is questionable whether the farmer in poorer countries can come up with the funds to "try out" and use the new crops. Another issue in this respect is how genetically engineered crops are perceived by the consumer. Even though in the US there is little resistance to such crops as long as the products can be shown to be safe and advantageous, in other countries genetically modified foods are received poorly by the consumer.

It is unlikely that there is a rationally sound basis for this rather hostile reaction of the consumer, as most of the crops are the results of human manipulation (such as centuries of breeding) and may have been treated with harmful herbicides and pesticides. Time and education will need to be invested to provide consumers and consumer advocates with a balanced opinion on the acceptability of the origin of their foods.

On the other hand, as so many plants (soybean, corn, etc.) are genetically modified and the nature of the genetic modification is not necessarily easy to explain, it may be simpler to label those foods that are guaranteed free of "genetically modified organisms" or their products. However, keep in mind that traditional breeding has genetically modified essentially all agricultural products, so it may be difficult to define what is actually free of genetically modified organisms.

Transgenic Crops

Transgenic crops: The varieties created through the use of a biotechnological approach commonly known as recombinant DNA, genetic engineering (GE), transgenic modification, or genetic modification (GM). The products of genetic engineering are often called genetically modified organisms, or GMOs. All these terms refer to methods of recombinant DNA technology by which biologists splice genes from one or more species into the DNA of crop plants to transfer chosen genetic traits.

Genes are segments of DNA that controlled the information that determines the structure and function of a living organism. Genetic engineers are molecular biologists who manipulate this information, typically by taking genes from one species—an animal, plant, bacterium, or virus—and inserting them into another species, such as an agricultural crop. With the advent of genetic engineering of plants around 1983, it appeared that this new biotechnology would benefit and even revolutionize agriculture.

The transfer of desirable genetic traits across species barriers has shown promise for solving problems in the management of agricultural crops. Potential benefits include reduced toxic pesticide use, improved weed control resulting in less tillage and soil erosion, and water conservation.

Gene Flow to Neighbouring Crops and to Related wild Species

Ecological scientists have little doubt that gene flow from transgenic fields into conventional crops and related wild plants will occur. Gene flow from transgenic to conventional crops is of concern to farmers because of its potential to cause herbicide resistance in related conventional crops.

For example, in western Canada, three different herbicide-resistant canola varieties have cross-pollinated to create canola plants that are resistant to all three types of herbicide. Through what is called gene-stacking, this new triple resistance has turned volunteer canola into a significant weed problem.

Gene flow from transgenic crops to wild relatives creates a potential for wild plants or weeds to acquire traits that improve their fitness, turning them into "super weeds." For example, if jointed goatgrass—a weedy relative of wheat—acquires the herbicide-tolerant trait of Roundup Ready wheat, it will thrive in crop fields unless applications of other herbicides are made. There is already evidence of such outcrossing from herbicide-resistant wheat to jointed goatgrass. Frank Young and his colleagues at Washington State University found that imidazolone-resistant wheat (not a transgenic variety) outcrossed to goatgrass in one season. Other traits that wild plants could acquire from transgenic plants that would increase their weediness are insect and virus resistance.

Because of their experience with classically bred plants, few scientists doubt that genes will move from crops into the wild: seven of the world's thirteen most important crop weeds have been made weedier by genes acquired from classically bred crops. Because gene flow has the potential to affect farmers' crop and pest management, crop marketability, and liability, more research needs to be done to determine the conditions under which gene flow from transgenic plants is likely to be significant.

Pesticide Resistance in Insect Pests

Bt has been widely used as a microbial spray because it is toxic only to caterpillars. In fact, it is a pest management tool that organic farmers depend on—one of the few insecticides acceptable under organic rules. Unlike the commercial insecticide spray, the Bt engineered into crop plants is reproduced in all, or nearly all, the cells of every plant, not just applied on the plant surface for a temporary toxic effect. As a result, the possibility that transgenic Bt crops will accelerate insect pests' development of resistance to Bt is a serious concern. Pest resistance to Bt would remove this valuable and environmentally benign tool from farmers' and forest managers' pest control toolbox. Antibiotic resistance

The use of antibiotic-resistant marker genes for the delivery of a gene package into a recipient plant carries the danger of spreading antibiotic-resistant bacteria. The likely result will be human and animal health diseases resistant to treatment with available antibiotics. Research is needed on antibiotic resistance management in transgenic crops. Already the European Commission's new rules governing transgenic crops stipulate phasing out antibiotic-resistant marker genes by the end of 2004.

Effects on Beneficial Organisms

Evidence is increasing that transgenic crops—either directly or through practices linked to their production—are detrimental to beneficial organisms. New studies are finding that Bt crops exude Bt in concentrations high enough to be toxic to some beneficial soil organisms. In addition, a study by University of Arkansas scientists has shown that "root development, nodulation and nitrogen fixation are impaired" in some varieties of Roundup Ready soybeans. The reason is that the beneficial rhizobium responsible for nitrogen fixation in soybeans is sensitive to Roundup. It also appears that disruption of beneficial soil organisms can interfere with plant uptake of phosphorus, an essential plant nutrient. Insecticidal crops in two ways can affect beneficial insects that prey on insect pests. First, the Bt in transgenic insecticidal crops has been shown in some laboratory studies to be toxic to ladybird beetles, lacewings, and monarch butterflies. The extent to which these beneficials are affected in the field is a matter of further study. Second, because the insecticidal properties of Bt crops function even in the absence of an economic threshold of pests, Bt crops potentially can reduce pest populations to the point that predator species are negatively affected.

Reduced Crop Genetic Diversity

As fewer and larger firms dominate the rapidly merging seed and biotechnology market, transgenic crops may continue the trend towards simplification of cropping systems by reducing the number and type of crops planted. In addition, seed saving, which promotes genetic diversity, is restricted for transgenic crops.

Food Safety

Because food safety is predominantly a concern of consumers, However, because food safety issues do impact marketing and international trade, here is a short list of consumer safety concerns about transgenic food:

- Possibility of toxins in food
- Possibility of new pathogens
- Reduced nutritional value

- Introduction of human allergens
- Transfer of antibiotic resistance to humans
- Unexpected immune-system and genetic effects from the introduction of novel compounds

Farm Management Issues

The most widely planted transgenic crops on the market today can simplify short-term pest management for farmers and ranchers. In the case of herbicide-tolerant crops, ideally farmers can use a single broad-spectrum herbicide for all their crop weeds, though they may need more than one application in a season. By planting insecticidal crops, farmers can eliminate the need to apply pesticides for caterpillar pests like the European corn borer or the cotton bollworm, though they still have to contend with other crop pests.

While these crops offer handy pest control features, they may complicate other areas of farm management. Farmers who are growing both transgenic and conventional varieties of the same crop will need to segregate the two during all production, harvesting, storage, and transportation phases if they are selling into differentiated markets or plan to save their own seed from their conventional crops.

To minimize the risk of gene flow from transgenic to adjacent conventional crop fields, federal regulations require buffer strips of conventional varieties around transgenic fields. Different transgenic crops require different buffer widths. Because the buffer strips must be managed conventionally, producers have to be willing to maintain two different farming systems on their transgenic fields. Crops harvested from the buffer strips must be handled and marketed as though they are transgenic. Planted refuges—where pest species can live outside fields of insecticidal and herbicide-tolerant transgenic crops—are also required to slow the development of weed and insect pest resistance to Bt and broad-spectrum herbicides. These refuges allow some individuals in the pest population to survive and carry on the traits of pesticide susceptibility. Requirements governing the size of refuges differ according to the type of transgenic crop grown.

Farmers growing herbicide-tolerant crops need to be aware that volunteer crop plants the following year will be herbicide-resistant. This herbicide resistance makes no-till or direct-seed systems difficult because volunteers can't be controlled with the same herbicide used on the rest of the crop. In a no-till system that relies on the same broad-spectrum herbicide that the volunteer plants are resistant to, these plants will contaminate the harvest of a following conventional variety of the same crop, a situation farmers need to avoid for two reasons. First, the contamination means a following conventional crop will have to be sold on the transgenic market. This leads to the second reason. If farmers grow and market a transgenic crop for which they do not have a technology agreement and did not pay royalty fees, they face possible prosecution by the company that owns the transgenic variety. Many farmers have been charged with "theft" of a company's patented seed as a result of contamination in the field. Farmers growing insecticidal crops need to recognize that insect pressure is difficult to predict and may not warrant the planting of an insecticidal variety every year. In a year when pest pressure is low, the transgenic seed becomes expensive insurance against the threat of insect damage.

Farmers growing transgenic crops need to communicate with their neighbours to avoid contaminating their fields and to ensure that buffers are adequate. In Maine, farmers growing transgenic crops are now required by law to be listed with the state agriculture department, to help identify possible sources of cross-contamination when it occurs. The new law, L.D. 1266, also "requires manufacturers or seed dealers of genetically engineered plants, plant parts or seeds to provide written instructions to all growers on how to plant, grow, and harvest the crops to minimize potential cross-contamination of non-genetically engineered crops or wild plant populations". Farm management issues common to all transgenic crops include yield, cost, price, profitability, management flexibility, sustainability, market acceptance, and liability.

Crop Yield, Costs, and Profitability

Some farmers will get higher yields with a particular transgenic crop variety than with their conventional varieties, and some will get lower yields. The same is true for costs. Some producers will see their overall profitability rise, and some will see it drop. If the results of research studies so far on yield, chemical use, costs, crop prices, and profitability of these crops seem contradictory, it is because farmers and farms are not all alike. Neither are the conventional varieties used for comparison. Some yield, cost, and profitability trends do appear to be emerging from the growing body of research data for transgenic crops, however. As noted in the Wallace Centre report, Roundup Ready soybeans were designed simply to resist a particular chemical herbicide, not to increase yields. In contrast, Bt corn and cotton, by resisting insect pests, may result in higher yields from reduced pest pressure.

Yield: Herbicide-tolerant Crops
—soybeans, Cotton, Canola

Herbicide-tolerant soybeans appear to suffer what's referred to as a "yield drag." Again, in some areas and on some farms this tendency of Roundup Ready soybean varieties to yield less than their comparable, conventional counterparts varies, but overall, they appear to average yields that are five to ten per cent lower per acre.. Drought conditions worsen the effects. The bacterium that facilitates nodulation and nitrogen fixation in the root zone apparently is sensitive to both Roundup and drought.

Yield: Insecticidal Crops—corn, Cotton

Bt cotton, especially, outpaces yields of conventional cotton by as much as 9 to 26 per cent in some cases, though not at all in others. Yield increases for Bt corn have not been as dramatic. Time will tell whether farmers can expect yield increases or decreases in the long run with these and other transgenic crop varieties.

Changes in Chemical Pesticide use

One of the promises of biotechnology is that it will reduce pesticide use and thereby provide environmental benefits and reduce farmers' costs. The herbicide-tolerant and insecticidal varieties are designed specifically to meet these goals. Increased future pesticide use resulting from the buildup of resistance to heavily used herbicides is a long-term concern. Pesticide use depends on the crop and its specific traits; weather; severity of pest infestations; farm management; geographic location of the farm; and other variables. As a result, conclusions drawn by various studies analysing pesticide use on transgenic crops remain controversial. According to the Wallace Centre report, in a review of the data available up through 2000, crops engineered to contain Bt appear to have decreased the overall use of insecticides slightly, while the use of herbicide-resistant crops has resulted in variable changes in overall herbicide use, with increases in use of some herbicides in some places and decreases in others

Profitability

Farmers need to consider all the factors that determine profitability. No single factor can tell the whole story. Transgenic crop seeds tend to be more costly, and farmers have the added expense of a substantial per-acre fee charged by the owners of transgenic varieties. These costs have to be considered along with input cost changes—whether herbicide or insecticide use and costs go down, go up, or stay the same. Market price is another factor: Prices for some transgenic crops in some markets are lower than prices for comparable conventional crops, though rarely are they higher.

Farmers need to watch the markets. Some buyers will pay a premium for a non-transgenic product, though as transgenic seeds find their way into conventional transportation, storage, and processing streams, these premiums may disappear along with confidence that "GMO-free" products are in fact truly free of engineered genes.

Marketing and Trade

Buyer acceptance is a significant marketing issue for farmers raising transgenic crops. Farmers need to know before they plant what their particular markets will or won't accept. Since most grain handlers cannot effectively segregate transgenic from non-transgenic crops in the same facility, many companies are channeling transgenic crops into particular warehouses. Farmers need to know which ones and how far away those are.

While the European Union officially lifted its two-year moratorium on the introduction of new transgenic crops in February 2001, during the debate over labeling and traceability regulations the moratorium remains in effect. Under the proposed new EU requirements, "all foods and animal feed derived from GMOs have to be labeled and, in the case of processed goods, records have to be kept throughout the production chain allowing the GMO to be traced back to the farm".

Liability

The need to separate transgenic crops from conventional and organic ones opens farmers to liability for their product at every step from seed to table. Effective systems for segregation do not exist at present, and will be costly to develop and put into place. Farmers may well end up bearing the added costs of crop segregation, traceability, and labeling. In the meantime, farmers who grow transgenic varieties—and, ironically, those who do not—are liable for transgenic seeds ending up where they aren't wanted: in their own non-transgenic crop fields, in neighbours' fields, in truckloads of grain arriving at the elevator, in processed food products on retail shelves, and in ships headed overseas.

Farmers who choose not to grow transgenic varieties risk finding transgenic plants in their fields anyway, as a result of cross-pollination via wind, insects, and birds, which may bring pollen from transgenic crops planted miles away. Besides pollen, sources of contamination include contaminated seed and seed brought in by passing trucks or wildlife. Those farmers whose conventional or organic crops are contaminated, regardless of the route, risk lawsuits filed against them by the companies that own the proprietary rights to seed the farmer didn't buy. Likewise, farmers who grow transgenic crops risk being sued by neighbours and buyers whose non-transgenic crops become contaminated.

Because contamination by transgenic material has become so prevalent in such a short time, all farmers in areas of transgenic crop production are at risk. Insurance, the most common recourse for minimizing potential losses because of liability, is not available to the nation's farmers for this risk because insurance companies do not have enough information to gauge the potential losses. A full risk assessment and legal clarification of the distribution of liability among farmers, seed companies, grain handlers, processors, and retailers is needed before farmers can rest assured that transgenic crops won't result in lawsuits against them.

For farmers who have adopted transgenic technology—and for those who have not—in the words of Brian Leahy, executive director of California Certified Organic Farmers, "This technology does not respect property rights". In essence, farmers who grow transgenic crops on some of their fields, and farmers who grow none, risk bearing tremendous liability. This situation won't change until farmers either gain legal protection or until they all grow transgenic crops on all their fields.

Influences on Public Research

While transgenic crop varieties are generally the property of private corporations, those corporations often contract with public-sector agricultural research institutions for some of their development work. In fact, private investment in agricultural research, including germplasm development, has surpassed public investment in recent years. With this shift in funding sources, the following questions become important: Is the private sector unduly influencing the public research agenda? Are corporations directing public research in less socially valuable directions while research on, for instance, sustainable agriculture goes wanting? Are the outcomes of corporate-funded transgenic research and development by our public institutions equitable across the food and agricultural sectors? Is equity even a consideration of our public institutions when they accept this work?

Because intellectual property rights (patents) apply to living organisms, making them private property, the free flow of scientific information that has historically characterized public agricultural research is being inhibited. What are the implications for the future of agriculture and society of the secrecy that now surrounds so much of what was formerly shared public knowledge? For a brief history of intellectual property rights as they apply to living organisms. These and other questions need to be addressed by citizens and their public institutions. These issues are of particular concern to farmers and consumers who would benefit from research into alternative technologies that are less costly (in every way), less risky, and more equitable.

Equity requires that the economic benefits and risks of transgenic technology be fairly distributed among technology providers, farmers, merchants, and consumers. For long-term sustainability, farmers need research that focuses on farms as systems, with internal elements whose relationships can be adjusted to achieve farm management goals. In contrast, transgenic crop research so far has focused on products that complement toxic chemical approaches to control of individual pest species. These are products that can be commercialized by large agribusiness or agri-chemical interests and that farmers must purchase every year. This research orientation only perpetuates the cost-price squeeze that continues to drive so many out of farming.

Industry Concentration and Farmers' Right to Save Seed

The broadening of intellectual property rights in 1980 to cover living organisms, including genes, has resulted in a flurry of mergers and acquisitions in the seed and biotech industries. According to the Wallace Centre report, "Relatively few firms control the vast majority of commercial transgenic crop technologies. These firms have strategically developed linkages among the biotechnology, seed, and agri-chemical sectors to capture as much market value as possible. However, these tightly controlled linkages of product sectors raise serious issues of market access, product innovation, and the flow of public benefits from transgenic crops. Unlike Plant Variety Protection—which does not allow for the patenting of individual genes, but only of crop varieties—Intellectual Property Rights prohibit farmers from saving their seed and undertaking their own breeding programme, and prohibit plant breeders from using the material to create a new generation of varieties adapted to specific regions or growing conditions.

Farmers can't save proprietary seed for planting and so must purchase new seed every year. In addition, farmers choosing transgenic varieties must sign a contract with the owner of the variety and pay a substantial per-acre technology fee, or royalty. The anticipated commercial introduction of transgenic wheat represents a dramatic shift in an industry in which farmers still widely save their own seed. As non-transgenic varieties become contaminated with transgenic ones, even those farmers who choose to stick with conventional varieties will lose the right to replant their own seed. This loss has already occurred in Canada's canola industry, with Monsanto winning its court case against farmer Percy Schmeiser for replanting his own canola variety that had become contaminated with Monsanto's Roundup Ready canola.

The adoption of transgenic crop varieties has brought with it an increasing prevalence of contract production. While contract production can lead to increased value and reduced risk for growers, farmers are justified in their concern about their loss of control when they sign a contract with a private company. Issues associated with contract production of transgenic crops must be considered within the broader context of a sustainable agriculture to include ownership, control, and social equity.

PLANT BREEDING MATERIAL AND COLLECTION

The advantage of having at one's disposal a large collection of plant varieties, both plants of commercial importance and imported plants of no immediate practical importance, is enormous. A difficult question for the plant breeder is always what crosses to make, what plants to combine with the greatest hope of ultimate success. In the breeding of such plants as potatoes, raspberries and roses, certain clones have the reputation of being profitable parents of hybrid lots. The Early Rose potato, the Lloyd George raspberry, the Chilean strawberry. In some special instances the choice is obvious. When our present assortment of flaxes or potatoes or roses happens to be deficient in some such quality as disease resistance or winter- hardiness, we are on the look-out for plants that will show those desirable qualities to the greatest extent.

In other words, we shall, when choosing the parents of our future hybrids, try to find material that promises the possibility of a combination of the very good points of both. This is obvious. Often crosses that seemed very little promising have given excellent results. It would seem that here no definite rule could be indicated. It is always worth while trying crosses with new material, with newly imported species—in fact with material which has not been used previously.

This is especially so when, as in horti- culture, everybody is on the look-out for something new. Another rule that could be given would be that when we make very wide crosses we cannot hope to find commercially valuable strains in the second hybrid generation, unless we grow extremely large numbers of plants in that generation. When we make a very wide cross to incorporate some special quality into the existent stocks, the best thing we can do is to use the hybrids in back-cross series. We can then strive to retain just one special quality of the new species and incorporate it into the general set of genes with which we are familiar in our older stock. In several instances very wide crosses have been successfully used, and sometimes plants have been used in cross-breeding that in themselves showed very little promise.

The inedible cattle gourd that Orton crossed with his water- melons, of some of the potato crosses with Andean wild species (Salaman) and of the back-crosses to hybrids with wild cane from which all our modern resistant varieties of sugar-cane have been derived. It is here—in the choosing of plant material for breeding work—that cytological studies and chromosomal counts are of the greatest practical value.

METHODS OF EVALUATION IN PLANT BREEDER

The aim of every plant breeder is to produce a group that will be worth a place in the assortment of cultivated plants. In some cases the new plant will have only ornamental value, in others it will, for any of a dozen reasons, be better adapted to agriculture than other similar ones. It is easy enough in all classes of plants to produce novelties, to produce variability. The experimental grounds of the plant breeder would soon be cluttered up with novelties if he did not keep on weeding out the comparatively worthless.

After producing our novel plants, we must judge them. Some qualities are much easier to judge than others. It is comparatively easy to weed out a lot of seedling dahlias and to pick out a few that may be worth keeping on for further testing. In the selection of our agricultural plants, in cereals, in beets, in maize, the assortment of valuable kinds is already so great that only relatively few of the newer selections are worth keeping.

In the vegetatively propagated plants this work of testing novelties is comparatively easy. In the first place, we are not bothered by considerations of purity—once we have produced a very good specimen, we can just multiply it with all its properties intact. In the second place, we can easily, for purposes of testing and comparison, produce as many individual plants of the novelty as we require. When the differences are relatively small—as in cereals and beets, where the novel plant produced must compete with excellent old stock that has been made by a process of selection and comparison during centuries—a correct system of evaluation becomes of very great importance. What we really want to test is the inherited composition of the material. This may be difficult, as the final quality of the plants to a large extent depends on favourable or unfavourable conditions under which they grew up.

It will perhaps be as well to begin with the simplest problems. In the production of novelties in such flowers as roses or dahlias the value of a new variety depends entirely on the appeal it makes to the gardening public. The man who is proud to have found a novel dahlia seedling or a new and strikingly beautiful rose sends it to a show and hopes that it will take the judge's eye and win him the much- "prized ribbon or certificate. But of course the insider knows that many highly commended novelties will never be heard of again, or that they will drop out of cultivation in a few years. There is a difference between the fact that the grower can bring half a dozen perfect blooms to the show on time, and the question whether that same rose or chrysanthemum will ever be popular in the gardens, or even be worth while growing for the commercial producer of cut flowers.

A dahlia that wins the medal may be a very weak grower, or it may be very susceptible to rain. The strikingly beautiful rose may be very susceptible to mildew, or its flowering season may be extremely short. If we visit the growers of greenhouse roses, or the gardens of the man who produces cut flowers for the market, we find that the number of varieties that have real commercial value—varieties "with bread in them ", as our gardeners express it—is extremely small. We shall probably find only three or four of the relatively older dahlias worth growing for cut flowers, and it is seldom that a novelty is added to that list. In testing out agricultural plants, both during the process of selection and after they have finally been introduced, many methods are in use.

The direct, simple examination of a number of small plots or rows will show us the qualities of those lots, but this method will only serve to find the groups with definitely inferior quality. It is not good enough to find exceptional merit, as circumstances may have favoured those qualities we appreciate. The only way to avoid reaching the wrong conclusions is to see to it that environmental conditions do not vitiate our results. The best way to do this is to com- pare the quality of a number of plants with that of other plants that were grown in wholly comparable conditions. One of the best ways to do this is to grow the plots of our plant between similar plots of an older variety, that can be used as a standard of comparison.

The choice of this standard is of great importance, and we will revert to this subject when treating of the breeding of special plant groups. As an example we can here take the evaluation of a newly produced (or of an imported) potato clone, or pure line of wheat or barley, In a case like this the standard may be some line or clone that is of commercial importance, against which the tested plant will be called upon to compete. If there is sufficient material, it is wise to repeat the comparison—that is to say, we should grow the test plot between standard plots in different parts of the field, or preferably in different fields.

To equalize accidental conditions in small parts of the field, oblong plots are generally better than square ones. If we are testing a number of experimental lots, we can grow them side by side, and interpolate standard plots to be used for purposes of comparison. The experimental garden will need an efficient system of book-keeping. We must note down anything of value about the experimental and the standard plots, date of maturity, disease resistance, quality, yield. We can calculate average yield for the sum of the plots of one kind, and then compare those figures with the average computed yield of the standard plots or with the average yield of the whole field; and a comparison of the deviation of the yields in the different plots with calculated mean errors will give us valuable data about the question in how far the superiority or inferiority of each separate number is significant as an indication of merit. The practice of plant breeding in different parts of the world.

A multitude of miserably small checkerboard fields being weighed at harvest time to produce the figures for an office staff to juggle with. A successful establishments where the field-work was under the supervision of men who were thoroughly conversant with the material, and where excellent results were obtained with a minimum of calculating machine work. A method of annotating the results obtained in experimental plots which gives the plant breeder an excellent insight into the merits of his new material is that of graphical registration. This method can be adapted to the most diverse projects; it is very simple, and gives us our results at a glance.

The method consists of registering the quality (yield, height, resistance) of each plot, both standard plots and testing plots, on ruled paper. If we do this, the line that can be drawn through the points of the standard plots will probably rise in one part of the field, and descend in some other part of it, and by drawing a line that joins those points we obtain a basis upon which we can found our judgment of the results obtained in the experimental plots. If we want to test our experimental material in two or three different qualities, we must construct separate graphs for each of those qualities, or sometimes we can use the same graph, employing different colours for separate qualities (weight, sugar percentage, purity in beets, etc.). Reading those graphs will now show us the quality of our plant plots, in comparison with the standards, and we can use this relative value as a basis for our selection work.

If we compare half a dozen groups in one experiment, and we find that one or two of them will in every duplication remain well below the standard in quality, this may be a very good reason for discarding them. On the other hand, if we find one or two numbers that all through the experiment are well above the standard line in quality, we have a very good reason for preferring them. It may be that in the case of one or two numbers the results are very variable, so that in one part of the field the number is above and in other parts it is below the standard in quality. Then we may well retain this number for another trial in the next season. The influence of one row of plants on those growing close to it may be considerable, and we must guard against this. If we can grow three rows and use only the middle one for our records, we take a step in the right direction.

If we have enough seed to plant six rows, and we examine only the two middle ones, discarding two rows on each side, this is still better. The greatest difficulty arises in those cases where we are dealing with very few plants of each number; or, worse, where, as in the second generation of a cross, every plant is probably different from every other. It then becomes very important to equalize conditions for each plant as far as this is possible. A very good plan is to grow the plants in rows, and alternate those experimental rows with rows of some pure standard variety. Each plant may then have two diverse neighbours in its own row, but it will touch four other plants that will be of the same kind as four out of six neighbour plants in the case of every other plant in the experiment.

This means that to save time we must make a preliminary choice out of the many numbers in those early generations, and it is very important that we shall make the cultural conditions as nearly equal as possible for every plant. This is the place to warn against using apparent correlative qualities in judging the value of our experimental plants. In the breeding of animals even to-day we often see that almost as much attention is paid to correlative qualities as to real merit. Some of this is certainly due to the fact that in some material real merit can only be judged in one of the two sexes. There is no good reason to give any attention to correlative qualities. In so far as a real correlation exists, selection for real quality will automatically bring along the correlative character, and in as far as we are mistaken (and this will nearly always be so) we are simply led astray by paying attention to apparent correlations.

In plant-breeding selection according to qualities believed to be indirect indications of good quality used to be very frequently attempted. This has gone out of fashion, and we only meet traces of it in very special fields (sugar-cane, apples). In our selection work we should always consider actual practical merit, and we should care- fully study in how far the qualities we can examine, measure and weigh are indications of this real merit.

GENETICALLY ENGINEERED PLANTS

At a symposium at the University of Washington Law School in October

1993, when discussing the 'Future of Intellectual Property Protection for Biotechnology in the US, EC, and Japan', Joseph Straus of the Max Planck Institute (Germany) warned the participants against 'ethics and other irrational considerations'.

Indeed, except for a very narrow band of 'moral dilemma' situations which are the stock-in-trade of professional biomedical ethicists, the ethical aspects of genetic engineering have been routinely ignored by government policy-makers and corporate technology promoters. My thesis is that it has been the corporate promoters and their governmental handmaidens who have been 'irrational' in their systematic refusal to acknowledge the environmental and ethical considerations of genetic manipulation.

Technologies, by definition, are neither acts of God nor nature; they are the embodiment of specific human purposes and intentionality. Far from being inevitable, they are researched and developed by those entities with sufficient power to mobilize social institutions to bring the technology into being, according to their own goals and normative considerations. 'Biotechnology is not neutral.

It shares the propensity of modern materialistic science to desacralise, dominate and manipulate life. It reduces all living things to a mechanism which it can manipulate according to engineering standards.' Genetic engineering is a technological process for undertaking activities which do not, and cannot, occur in nature. In this sense it is perfectly legitimate, therefore, to label genetic engineering as 'unnatural'. The European Community has incorporated this concept in its definition of genetically modified organisms, and it has been adopted by the UN Environment Programm (in the working of its Fourth Expert Panel set up under the Biodiversity Convention as part of the follow through to the United Nations Conference on the Environment and Development held in Rio de Janeiro in June 1992): 'organisms in which the genetic material has been altered in a way that does not occur naturally by mating and/or natural recombination'.

Thus, hybrids and other modified organisms obtained by traditional breeding techniques are excluded from the definition of genetically modified organisms. It is important to note that the process as well as the products are novel. The promoters of the new technology believe that a concern for the process itself is specious, downplaying its novelty whenever confronted by discussions of regulatory oversight. However, one of the bases of the decision of the US Supreme Court in the *Chakrabarti* case (which, 5 to 4, upheld the patentability of genetically engineered microbes, and which has been unquestionably accepted by foreign governments as disposative of the issue of patentability of genetically engineered life forms) was that the element of 'novelty' required by the patent law was to be found in the process by which the microbe had been created. One reason why we cannot ignore the powerful novel aspects of the processes of genetic manipulation is because we are not omniscient as to what will, in fact, happen when we alter genomes. Genetic manipulation is not like a child's game of Legos or Tinkertoy, in which parts can be rearranged or linked up with only simple mechanical and relatively predictable consequences. In calculating any risk from a transgenic organism, one should consider four elements: the host organism, the foreign genes, the interaction between the foreign genes and the rest of the genome, and the environment in which the organisms will be used.... [in regard to the last two elements] the literature contains many examples of genetic manipulations where inserted genes did not respond in their new environments the way they did in their old ones or where alterations with one part of the genome caused surprising activity in other parts of the genome. There is an element of arrogance to scientific assertions which assure us that the process itself poses no risks, since scientists know so little about actual ecosystems.

For example, I have been told by agronomists that over 80% of the organisms which can be identified in a soil sample from my garden are completely unknown to the scientific literature. Indeed, the Ecological Society of America itself has specifically warned about the problems presented in our lack of ecological knowledge and the resulting consequences from releasing (intentionally or accidentally) genetically altered organisms into ecological systems.

Foods

The Food and Drug Administration, responding to corporate pressures to remove the prospect of regulation of genetically altered foods, issued a set of rules which have largely left responsibility for protecting the public health and safety in the hands of the industry, permitting products to be marketed without scrutiny unless the industry indicated to the agency that it believed governmental oversight was justifiable. However, the proposal as published by the FDA in the Federal Register perversely noted several important problem areas: creating allergens, additions of genes from sources which might violate religious and cultural norms (of vegetarians, Jews or Muslims, etc.), and implications for animal welfare and well-being (for example, the incorporation of human growth hormone-producing gene into a pig's genome produced a highly arthritic animal).

Indeed, the commissioner of the FDA, David A. Kessler, and his colleagues noted the possibility that genetically engineered food might 'contain high levels of unexpected, acutely toxic substances'. In order to begin to address ethical issues presented by the genetic modification of foodstuffs, we need to have some ideas of the purposes for which these modifications are performed. (For example, if we wish to approach an analysis on utilitarian cost-benefit grounds.)

Despite a great number of general statements, human nutrition and hunger do not appear to be the actual driving forces behind the development of genetically engineered foods. Other than claiming increased shelf life (which can be indirectly related to nutrition in the sense of reducing spoilage or the consumption of foods that have begun to spoil), there is hardly any indication at all that genetic engineering is being directed to create nutritious substances out of nonnutritious ones.

There are few real instances of genetic engineering reducing the cost of foodstuffs, or increasing the quantities of foods actually available to populations that are hungry because of the non-existence of consumables (as opposed to hungry because they lack money to buy sufficient foods, no matter how produced). Indeed, early genetic manipulations seemed perversely designed to minimize the achievement of such goals: the creation of 'herbicide tolerant plants' which do not reduce the applications of dangerous chemicals to agricultural fields (which might occur if genetic manipulations of food crops were designed to make them resistant to insects, fungi, or disease) but instead permit higher levels of chemical application, or the introduction of recombinant bovine growth hormone to produce more milk at a time when developed countries suffer from milk gluts and have instituted programmes to kill cows and physically dump milk (milk as a commodity is price inelastic; quantity increases do not lead to price decreases).

The goals of genetically engineering foods seem to be more closely tied to increasing the economic gain and power of the corporations involved in food production (for example, the development of herbicide-tolerant farm crops has been led by corporations which manufacture herbicides), making production easier for corporations, reducing or altering packaging and transportation costs for the corporations, etc.

Environment

Industrial societies around the world are faced by massive pollution legacies from the previous emphases on chemical and nuclear industrial activities. The inability of the US Department of Energy to find a suitable site for a longterm nuclear waste repository illustrates our very poor track record for dealing with the scientific, technical, economic, and socio-political aspects of prior technological 'revolutions'.

The environmental problems posed by genetically engineered organisms are likely to be substantially more intractable than those posed by these earlier instances of pollution because genetic wastes multiply, migrate, and mutate. A genetically engineered organism once free in the environment is impossible to recall. (Impacts which are irreversible must be considered with higher scrutiny than those which can be undone.) Examples of environmental problems which need to be assessed include the risks of transgenic crops themselves becoming weeds; the risk of gene flow to wild relatives which might become weeds or pests; the growth of antibiotic resistance in species (particularly animals), since resistance genes are used as markers for the genetic engineering sites; the problem of exotic or non-native species taking over ecological niches (such as gypsy moth, starling, kudzu, rabbits in Australia, etc.) leading to extraordinary economic losses as well as ecological ones.

The restriction (rather than increase) of biodiversity by selective advantageous breeding of transgenic organisms as opposed to natural ones, or the predatory results of expanding exotic species (such as the Dutch Elm micro-organism destroying the American Elm and severely restricting the biodiversity of certain areas in the Northeast United States); health issues (both of plant and animal species, as well as humans — worker health and safety as well as community security); and the occurrence of the completely unexpected, the inability to eliminate uncertainty

For example, the late 1993 floods in the Mississippi valley included the flooding of a field of genetically engineered corn and the dispersal of this plant material to unknown sites within the thousands of square miles of downstream flood plain. Prospective ecological assessment is not being performed, despite the fact that its need has long since been recognized. And economic priorities, particularly those accruing directly to the promoters, as well as the nebulous spur of 'competition' with foreign countries, oftentimes overwhelms any interest in doing environmental impact analysis.

BIOETHICS IN PLANT GENETIC ENGINEERING

There are issues and concerns regarding the use of transgenic crops and their effects on the health and the environment in general.

The major concerns about GM crops and GM foods are:

- Effect of GM crops on biodiversity and environment- As the GM crops are created artificially, there is no natural process of evolution in their development. Hence, there is a question of this affecting the biodiversity and overall effect on the environment.
- The risk of transfer of transgene from GM crops to pathogenic microbes- Antibiotic marker genes are used to identify and select the modified cells. If GM food containing antibiotic resistance marker gene is consumed by animals and humans, there is a risk that the transgene will transfer from GM food to microflora of human and animals. This may lead to the gut microbes to become resistant to antibiotics.
- The transfer of genes from animals into Gm crops for molecular farming may change the fundamental vegetable nature of plants.
- The GM crops may bring about changes in evolutionary patterns. The plants adapt to the changing environment in the natural way by changing their genes and developing better races with superior traits which ultimately leads to the development of evolved races and varieties. What will be the evolutionary pattern of the GM crops?

There are concerns about the effect of transgene flow from GM crops to other non-GM plants and the alteration of these non-GM crops.
- There is a risk of transferring allergens (usually glycoproteins) from GM food to human and animals.
- There is a risk of "gene pollution" *i.e.* transfer of transgene of GM crop through pollen grains to related plant species and development of super weeds.
- There are also some religious issues related to the consumption of transgenic plants with animal genes introduced into them, especially, for some strict vegetarian people and some ethnic groups with certain food preferences and restrictions.
- There is a need to study thoroughly as to how the genetically engineered plants will affect the ecological balance, once they are released in the environment.

TISSUE CULTURE TECHNIQUES

An important aspect of all biotechnology processes is the culture of either the plant cells or animal cells or microorganisms. The cells in culture can be used for recombinant DNA technology, genetic manipulations etc.

Plant cell culture is based on the unique property of the cell-totipotency. CELL-TOTIPOTENCY is the ability of the plant cell to regenerate into whole plant. This property of the plant cells has been exploited to regenerate plant cells under the laboratory conditions using artificial nutrient mediums. With the advances made in genetic engineering, it became possible to introduce foreign genes into cell and tissue culture systems. This led to the development of Genetically Modified (Gm) or Transgenic Crops which had improved traits and characteristics.

HISTORY OF CELL CULTURE

In the early 19th century, Schleiden and Schwann proposed the concept of the 'cell theory'. In 1902, Gottlieb Haberlandt, the german botanist and regarded as the father of plant tissue culture, first attempted to cultivate the mechanically isolated plant leaf cells on a simple nutrient medium. He did not succeed in achieving the growth and differentiation of the cultured cells, however, he predicted the concept of growth hormones, the use of embryo sac fluids, the cultivation of artificial embryos from somatic cells, etc.

During the period 1902 - 1930, attempts were made to culture the isolated plant organs such as roots and shoot apices (organ culture). Hanning isolated embryos of some crucifers and successfully grew on mineral salts and sugar solutions. Simon successfully regenerated a bulky callus, buds, roots from a poplar tree on the surface of medium containing IAA which proliferated cell division. Gautheret, White and Nobecourt largely contributed to the developments made in plant tissue culture. White cultured tobacco tumour tissue from the hybrid Nicotiana glauca, and N. Langsdorffii. The period saw the development of suitable nutrient media to culture plant tissues, embryos, anthers, pollen, cells and protoplasts, and the regeneration of complete plants (in vitro morphogenesis) from cultured tissues and cells. In 1941, van Overbek and co-workers used coconut milk (embryo sac fluid) for embryo development and callus formation in Datura. Steward and Reinert first discovered somatic embryo production in vitro. Maheswari and Guha developed the anther culture for the production of haplid plants. Skoog and Miller advanced the hypothesis of organogenesis in cultured callus by varying the ratio of auxin and cytokinin in the growth medium. Muir developed a successful technique for the culture of single isolated cells wich is commonly known as paper-raft nurse technique (placing a single cell on filter paper kept on an actively growing nurse tissue). In 1952, the Pfizer Inc., New York got the US patent and started producing industrially the secondary metabolites of plants. The first commercial production of a natural product shikonin by cell suspension culture was obtained.

In 1980s using Genetic engineering, for the first time, it was possible to introduce foreign genes into cell and tissue culture systems to develop plants with improved characteristics (transgenic crops) which may contribute to the path towards the second green revolution.

Table. List of Transgenic Animals Produced With Their Promoter, Enhancer and Structural Transgenes

Animal	**Genes Tansferred (Promoter or Enhancer/Structural Transgenes)**
Goat	A variant of tPA gene (LAtPA)
Sheep	mMT/hGH, mMT/TK, mMT/bGH, mMT/hGRF, oBLG/hFIX, oBLG/alpha1AT, oMT/oGH
Rabbit	mMT/hGH, hMT/hGH, rbEu/rb,c-myc
Pig	mMT/hGH, mMT/bGH, hMT/pGH, MLV/rGH, MLV/rGH, bPRL/bGH
Fish	hGH, mMT/hGh, mMT/bGal, cd-crystallin SV/hygro, AFP
Cow	BPV, lactoferrin
Chicken	ALV, REV
Mouse	mMT/rGH, mMT/bGH, mMT/oGH, mMT/hGh, mMT/hGRF, mMT/hFIX

A transgenic animal is an individual in which a gene (or genes) from another individual has been artificially inserted with the new genetic information forming a permanent part of the genome and being passed on to their offspring according to the Mendelian laws of inheritance. So far, transgenic cattle, sheep, pigs, rabbits, and chicken have been produced. After fertilization a mammalian embryo passes through many stages of development such as formation of pronuclei, morula formation, blastocyst formation, implantation etc. Using these early development stages of mammalian embryos, genetic engineers have used different methods to introduce foreign genes into the genome.

METHODS TO INDUCE FOREIGN GENES INTO THE GENOME

Micro-injection in to a Pronucleus

A fertilized egg is held in a pipette by suction and several copies of the foreign genes are injected into one of the pronuclei via a micropipette. Although this has been the most widely used and most successful method, 30% of the treated embryos degenerate and die with in a few hours.

Using Retroviruses as Vectors to Carry Foreign DNA into Morulas

A morula is placed on a culture of fibroblasts infected with retrovirus after removing the zona pellucida layer. The virus cannot penetrate this layer therefore it is essential to get rid of it. The retrovirus, genetically engineered to carry foreign DNA, infect the cells of the morula as they are shed from the fibroblast. This technique often creates 'mosaic offspring', where some cells contain the foreign gene while others do not.

One of the limitations of this technique is that the modified virus cannot leave the transgenic species and under certain conditions it could mutate regaining its ability to cause disease in the tissues where it occurs.

Retroviral Infection of the Stem cells, which are then Injected into the Cavity of Another Blastocyst

In this method, stem cells, from the inner cell mass of an embryo, are infected with genetically-engineered retroviruses, then injected into the central cavity of a different blastocyst. The injected cells colonise the new embryo and participate in the formation of all the tissues, including the ovaries and testes from which cells are formed. This technique has been used to produce transgenic mice and hamsters.

The first transgenic animals produced were mice, into which a gene cloning for rat growth hormone was inserted by microinjection.

Two techniques were used to introduce the new gene into the mouse genome:

- As many as 10000 copies of the rat gene were injected into a pronucleus of a fertilized mouse egg, using a microneedle.
- After treatment with calcium chloride, to make them more permeable, the rat DNA was applied to the outside of a 2-8- celled mouse embryo in culture.

The treated eggs, or embryos were then transplanted into foster mothers. Embryos with the inserted genes were identified by DNA hybridization with small samples of DNA from white blood cells or skin cells. Transformed adult mice produced the growth hormone mainly in their livers rather than pituitary gland where it is normally synthesized.

This technique can be used successfully to transfer other genes such as those affecting the properties of hair, hides, cold tolerance, disease-resistance and milk production which might help to produce strains of farm animals with economic advantage over existing breeds and stocks. In this regard, research is going on to use techniques which can improve the quality of the farm animals. In this regard, following two techniques are of great importance.

IMPROVING THE QUALITY

EMBRYO MANIPULATION

In this method, the fertilized egg is bisected at the two celled stage and each half is transplanted into different regions of the uterus. Using this method the reproductive rate is doubled as we know that female sheep and cattle produce, on an average, one offspring per pregnancy. Hence the farmers can easily increase their farmstock.

This technique can also be used to conserve rare breeds. After fertilizing their eggs in the laboratory , the young embryos or rare animals are dissected into anything from 2-8 cells. Each cell, is then transplanted into a surrogate mother of a common breed where it grows to produce a new individual of the rare type.

Another variation of this technique allows surrogate mothers to carry embryos of a different species *e.g.* Horses giving birth to zebras from zebra embryos implanted at the blastocyst stage. The outer layer of zebra embryo is exchanged with the trophectoderm of the horse embryo in order to make it acceptable to the surrogate mother. This exchange of the layer does not affect the development of the inner cell mass which is the true embryo-forming region.

EMBRYO CLONING

Using this technique it is possible to produce many genetically identical copies of an animal. This method has been used to clone mouse however, research is going on to standardize this method to obtain clones of cattle with desirable traits *e.g.* cows with high milk production etc.

Following steps are used in this method:

- An egg from the donor is grown to blastocyst stage under laboratory conditions.
- The egg is dissected to remove the inner cell mass.
- The mass cells are separated into individual cells.

- Injection of a nucleus from each of these cells into a one-celled embryo containing two pronuclei.
- Removal of pronuclei followed by culturing of embryos in the laboratory until the blastocyst stage.
- Transplantation of these embryos into surrogate mothers.

CHIMERAS

Early research into the production of transgenic animals revealed a simple method for producing hybrids, called Chimeras, between closely related species. Sheep-goat chimeras were produced by mixing four celled sheep embryos with eight-celled goat embryos. After removing the zona pellucida from each egg, the eggs were pressed together and incubated at 370C. The cells reorganized and formed hybrid blastocysts. These hybrid blastocysts were then transferred to sheep foster mothers, where they continued their growth and development until the sheep gave birth. Each hybrid offspring contained both sheep and goat cells in all it's tissues which resulted in the coats of these animal having a patchwork kind of pattern with irregular patches of sheep and goat fur.

ENVIRONMENTAL TECHNIQUES

The term "Environment" is defined as our surroundings which includes the abiotic component (the non living) and biotic component (the living) around us. The abiotic environment includes water, air and soil while the biotic environment consists of all living organisms – plants, animals and microorganisms. Environmental pollution broadly refers to the presence of undesirable substances in the environment which are harmful to man and other organisms. In the past decade or two, there has been a significant increase in the levels of environmental pollution mostly due to direct or indirect human activities. The major sources of environmental pollution are –Industries, Agricultural sources (mainly rural area), anthropogenic sources (man related activities mainly in urban areas), biogenic sources etc. The pollutants are chemical, biological and physical in nature. The Chemical pollutants include-gaseous pollutants (hazardous gases like sulfur dioxide, nitrogen oxide), toxic metals, pesticides, herbicides toxins and carcinogens

Etc. The physical pollutants are- heat, sound, radiation, and radioactive substances. The pathogenic organisms and some poisonous and dangerous biological products are the biological pollutants.

Controlling the environmental pollution and the conservation of environment and biodiversity and controlling environmental pollution are the major focus areas of all the countries around the world. In this context the importance and impact of biotechnological approaches and the implications of biotechnology has to be thoroughly evaluated. There have been serious concerns regarding the use of biotechnology products and the impact assessment of these products due to their interaction with the environmental factors. A lobby of the environmentalists have expressed alarm on the release of genetically engineered organisms in the atmosphere and have stressed on thorough investigation and proper risk assessment of theses organisms before releasing them in to the environment. The effect of the effluents from biotechnological companies is also a cause of concern for everyone. The need of the hour is to have a proper debate on the safety of the use of the biotechnological products.

The efforts are not only on to use biotechnology to protect the environment from pollutionbut also to use it to conserve the natural resources. As we all know that microorganisms are known natural scavengers so the microbial preparations (both natural as well as genetically engineered) can be used to clean up the environmental hazards.

DEVELOPMENT OF ALTERNATE CLEANER TECHNOLOGIES USING BIOTECHNOLOGY

Biotechnology is being used to provide alternative cleaner technologies which will help to further reduce the hazardous environmental implications of the traditional technologies. *E.g.* some Fermentation technologies have some serious environmental implications. Various biotechnological processes have been devised in which all nutrients introduced for fermentation are retained in the final product, which ensures high conversion efficiency and low environmental impact.

In paper industry, the pulp bleaching technologies are being replaced by more environmentally friendly technologies involving biotechnology.

The pulp processing helps to remove the lignin without damaging valuable cellulosic fibres but the available techniques suffer from the disadvantages of high costs, high energy use and corrosion. A lignin degrading and modifying enzyme (LDM) was isolated from Phanerochaete chrysosporum and was used, which on one hand, helped to reduce the energy costs and corrosion and on the other hand increased the life of the system. This approach helped in reducing the environmental hazards associated with bleach plant effluents.

In Plastic industry, the conventional technologies use oil based raw materials to extract ethylene and propylene which are converted to alkene oxides and then polymerized to form plastics such as polypropylene and polyethylene.

There is always the risk of these raw materials escaping into the atmosphere thereby causing pollution. Using biotechnology, more safer raw materials like sugars (glucose) are being used which are enzymatically or through the direct use of microbes converted into alkene oxides.*e.g.* Methylococcus capsulatus has been used for converting alkene into alkene oxides.

BIOREMEDIATION

Bioremediation is defined as 'the process of using microorganisms to remove the environmental pollutants where microbes serve as scavengers. The removal of organic wastes by microbes leads to environmental cleanup. The other names/terms used for bioremediation are biotreatment, bioreclamation, and biorestoration. The term "Xenobiotics" (xenos means foreign) refers to the unnatural, foreign and synthetic chemicals such as pesticides, herbicides, refrigerants, solvents and other organic compounds. The microbial degradation of xenobiotics also helps in reducing the environmental pollution.

Pseudomonas which is a soil microorganism effectively degrades xenobiotics. Different strains of Pseudomonas that are capable of detoxifying more than 100 organic compounds (*e.g.* phenols, biphenyls, organophosphates, naphthalene etc.) have been identified. Some other microbial strains are also known to have the capacity to degrade xenobiotics such as Mycobacterium, Alcaligenes, Norcardia etc.

FACTORS AFFECTING BIODEGRADATION

The factors that affect the biodegradation are: the chemical nature of xenobiotics, the concentration and supply of nutrients and O_2, temperature, pH, redox potential and the capability of the individual microorganism. The chemical nature of xenobiotics is very important because it was found out that the presence of halogens *e.g.* in aromatic compounds inhibits biodegradation.

The water soluble compounds are more easily degradable whereas the presence of cyclic ring structure and the length chains or branches decrease the efficiency of biodegradation. The aliphatic compounds are more easily degraded than the aromatic ones.

BIOSTIMULATION

It is a process by which the microbial activity can be enhanced by increased supply of nutrients or by addition of certain stimulating agents like electron acceptors, surfactants etc.

BIOAUGMENTATION

It is possible to increase biodegradation through manipulation of genes *i.e.* using genetically engineered microorganisms and by using a range of microorganisms in biodegradation reaction.

Depending on the method followed to clean up the environment, the bioremediation is carried out in two ways:

In Situ Bioremediation

In situ bioremediation involves a direct approach for the microbial degradation of xenobiotics at the site of pollution which could be soil, water etc. The adequate amount of essential nutrients is supplied at the site which promotes the microbial growth at the site itself. The in situ bioremediation is generally used for clean up of oil spillages, beaches etc.

There are two types of in situ bioremediation:

1. *Intrinsic bioremediation*: The microorganisms which are used for biodegradation are tested for the natural capability to bring about biodegradation. So the inherent metabolic ability of the microorganisms to degrade certain pollutants is the intrinsic bioremediation. The ability of surface bacteria to degrade a given mixture of pollutants in ground water is dependent on the type and concentration of compounds, electron acceptor and the duration of bacteria exposed to contamination. Therefore, the ability of indigenous bacteria degrading contaminants can be determined I laboratory by using the techniques of plate count and microcosm studies. The conditions of site that favour intrinsic bioremediation are ground water flow throughout the year carbonate minerals to buffer acidity produced during biodegradation, supply of electron acceptors and nutrients for microbial growth and absence of toxic compounds.
2. *Engineered in situ bioremediation*: When the bioremediation process is engineered to increase the metabolic degradation efficiency (of pollutants) it is called engineered in situ bioremediation. This is done by supplying sufficient amount of nutrients and oxygen supply, adding electron acceptors and maintaining optimal temperature and pH. This is done to overcome the slow and limited bioremediation capability of microorganisms.

Advantages of in Situ Bioremediation

- The method ensures minimal exposure to public or site personnels.
- There is limited or minimal disruption to the site of bioremediation.
- Due to these factors it is cost effective.
- The simultaneous treatment of contaminated soil and water is possible.

Disadvantages of in Situ Bioremediation

- The sites are directly exposed to environmental factors like temperature, oxygen supply etc.
- The seasonal variation of microbial activity exists.
- Problematic application of treatment additives like nutrients, surfactants, oxygen etc.
- It is a very tedious and time consuming process.

Ex-situ Bioremediation

In this the waste and the toxic material is collected from the polluted sites and the selected range of microorganisms carry out the bioremediation at designed place. This process is an improved method over the in situ bioremediation method.

On the basis of phases of contaminated materials under treatment ex-situ bioremediation is classified into two:

1. *Solid phase treatment*: This system includes land treatment and soil piles comprising of organic wastes like leaves, animal manures, agricultural wastes, domestic and industrial wastes, sewage sludge, and municipal solid wastes. The traditional clean-up practice involves the informal processing of the organic materials and production of composts which may be used as soil amendment. Composting is a self heating, substrate-dense, managed microbial system which is used to treat large amount of contaminated solid material. Composting can be done in open system *i.e.* land treatment and/or in closed treatment system. The hazardous compounds reported to disappear through composting includes aliphatic and aromatic hydrocarbons and certain halogenated compounds. The possible routes leading to the disappearance of hazardous compounds include volatilization, assimilation, adsorption, polymerization and leaching.
2. *Slurry phase treatment*: This is a triphasic treatment system involving three major components- water, suspended particulate matter and air. Here water serves as suspending medium where nutrients, trace elements, pH adjustment chemicals and desorbed contaminants are dissolved. Suspended particulate matter includes a biologically inert substratum consisting of contaminants and biomass attached to soil matrix or free in suspending medium. The contaminated solid materials, microorganisms and water formulated into slurry are brought within a bioreactor *i.e.* fermenter. Biologically there are three types of slurry-phase bioreactors: aerated lagoons, low shear airlift reactor, and fluidized-bed soil reactor. The first two types are in use of full scale bioremediation, while the third one is in developmental stage.

Advantages of Ex-situ Bioremediation

- As the time required is short, it is a more efficient process.
- It can be controlled in a much better way.
- The process can be improved by enrichment with desired and more efficient microorganisms.

Disadvantages of Ex-situ Bioremediation

- The sites of pollution remain highly disturbed.
- Once the process is complete, the degraded waste disposal becomes a major problem.
- It is a costly process.

Several types of reactions occur during the bioremediation/microbial degradation:

- *Aerobic bioremediation*: When the biodegradation requires oxygen O_2 for the oxidation of organic compounds, it is called aerobic bioremediation. Enzymes like monooxygenases and dioxygenases are involved and act on aliphatic and aromatic compounds.
- *Anaerobic bioremediation*:This does not require oxygen O_2. the degradation process is slow but more cost effective since continuous supply of oxygen is not required.
- *Sequential bioremediation*: Some of the xenobiotic degradation requires both aerobic as well as anaerobic processes which very effectively reduces the toxicity *e.g.* tetrachloromethane and tetrachloroethane undergo sequential degradation.

Use of genetic engineering and genetic manipulations for more efficient bioremediation. In recent years, efforts have been made to create genetically engineered microorganisms (GEMs) to enhance bioremediation. This is done to overcome some of the limitations and problems in bioremediation.

These problems are:

- Sometimes the growth of microorganisms gets inhibited or reduced by the xenobiotics.
- No single naturally occurring microorganisms has the capability of degrading all the xenobiotics present in the environmental pollution.
- The microbial degradation is a very slow process.
- Sometimes certain xenobiotics get adsorbed on to the particulate matter of soil and thus become unavailable for microbial degradation.

As the majority of genes responsible for the synthesis of enzymes with biodegradation capability are located on the plasmids, the genetic manipulations of plasmids can lead to the creation of new strains of bacteria with different degradative pathways. In 1970s, Chakrabarty and his team of co-workers reported the development of a new strain of bacterium Pseudomonas by manipulations of plasmid transfer which they named as "superbug". This superbug had the capability of degrading a number of hydrocarbons of petroleum simultaneouslysuch as camphor, octane, xylene, naphthalene etc. In 1980, United States granted the patent to this superbug making it the first genetically engineered microorganism to be patented.

In certain cases, the process of plasmid transfer was used. *E.g.* The bacterium containing CAM (camphor degrading) plasmid was conjugated with another bacterium with OCT (octane degrading) plasmid. Due to non-compatibility, these plasmids cannot coexist in the same bacterium. However, due to the presence of homologous regions of DNA, recombination occurs between these two plasmids which results in a single CAM-OCT plasmid giving the bacterium the capacity to degrade both camphor as well as octane. A new strain of Pseudomonas sp. (strain ATCC 1915) has been developed for the degradation of vanillate (which is a waste product from paper industry) and sodium dodecyl sulfate (SDS, a compound used in detergents).

BIOTECHNOLOGICAL METHOD TO REDUCE ATMOSPHERIC CARBON DIOXIDE (CO_2)

Carbon dioxide is the gas that is the main cause of green house effect and rise in the atmospheric temperature. During the past 100-150 years, the level of CO_2 has increased about 25% with an increase in the atmospheric temperature by about 0.5% which is a clear indication that CO_2 is closely linked with global warming. There is a steady increase in the CO_2 content due to continuous addition of CO_2 from various sources particularly from industrial processes. It is very clear that the reduction in atmospheric CO_2 concentration assumes significance.

Biotechnological methods have been used to reduce the atmospheric CO_2 content at two levels:

1. *Photosynthesis*: Plants utilize CO_2 during the photosynthesis which reduces the CO_2 content in the atmosphere. The equation for photosynthesis is:
   sunlight
   $$6CO_2 + 6H_2O \rightarrow C_6H_{12}O_6 + 6O_2$$
   Chlorophyll

   The fast growing plants utilize the CO2 more efficiently for photosynthesis. The techniques of micropropagation and synthetic seeds should be used to increase the propagation of such fast growing plants.

   Further, the CO_2 utilization can be increased by enhancing the rate of photosynthesis. The enzyme ribulose biphosphate carboxylase (RUBP-case) is closely linked with CO_2 fixation. The attempts are being made to genetically manipulate this enzyme so that the photosynthetic efficiency is increased.

   Some microalgae like Chlorella pyrenodiosa, Spirulina maxima are known to be more efficient than higher plants in utilizing atmospheric CO_2 for photosynthesis and generate more O_2 than the amount of CO_2 consumed.

   The growing of these microalgae near the industries and power plants (where the CO_2 emission in to atmosphere is very high) will help in the reduction of polluting effects of CO_2. Using genetic engineering, attempts are going on to develop new strains of these microalgae that can tolerate high concentrations of CO_2. A limited success has already been reported in the mutants of Anacystis nidulans and Oocystis sp.
2. *Biological Calcification*: Certain deep sea organisms like corals, green and red algae store CO_2 through a process of biological calcification.

As the $CaCO_3$ gets precipitated, more and more atmospheric CO_2 can be utilized for its formation. The process of calcification is as follows:

$$H_2O + CO_2 \rightarrow H_2CO_3$$
$$H_2CO_3 + CaCO_2 \rightarrow CaCO_3 + CO_2 + H_2O$$

TREATMENT OF SEWAGE USING MICROORGANISMS

The sewage is defined as the waste water resulting from the various human activities, agriculture and industries and mainly contains organic and inorganic compounds, toxic substances, heavy metals and pathogenic organisms. The sewage is treated to get rid of these undesirable substances by subjecting the organic matter to biodegradation by microorganisms. The biodegradation involves the degradation of organic matter to smaller molecules (CO_2, NH_3, PO_4 etc.) and requires constant supply of oxygen. The process of supplying oxygen is expensive, tedious, and requires a lot of expertise and manpower. These problems are overcome by growing microalgae in the ponds and tanks where sewage treatment is carried out. The algae release the O_2 while carrying out the photosynthesis which ensures a continuous supply of oxygen for biodegradation.

THE SCIENCE OF MODERN GENETIC ENGINEERING

Genetic engineering is one form of biotechnology and usually refers to copying a gene from one living organism—plant, animal or microbe—and adding it to another organism. In genetic engineering, a small piece of genetic material (DNA) is inserted into another organism to produce a desired effect. This is in contrast to traditional plant breeding in which all the genes desirable and undesirable contained in the male plant—pollen—are combined with all the genes of the female plant.

The progeny resulting from this cross may contain the gene for the desirable character, but it will also contain many of the undesirable genes from both parents. Genetic engineering has the advantage of being able to transfer only the gene of interest and greatly accelerate plant breeding. But genetic engineering also is more powerful than traditional breeding since it can move genes not only between similar plant species but also from distant relatives, including non-plant species. It is possible to move genes between such seemingly unrelated organisms because all living organisms share the same code for DNA and the synthesis of proteins and other basic life functions. What might seem on the surface to be very different organisms are, in fact, very similar, at least at the molecular level. All living things are more alike than different, and this is one of the reasons that genes can be moved so successfully between such seemingly different organisms as plants and bacteria. Genes are not unique to the organisms from which they came, so there really aren't "tomato genes" or "bacterial genes." It's the collection of all genes in a tomato or a bacterium that makes it a tomato or bacterium, not a single gene. As we learn more about the genetic makeup of all organisms, we see that most plant species differ by only a small percentage of their genes and those even such seemingly different organisms as tomatoes and bacteria have many common genes.

These findings suggest that in the long-term evolutionary process even tomatoes and bacteria had some common ancestor. From the discovery 50 years ago of the structure of DNA, scientists soon came to realise they could take segments of DNA that carried information for specific traits—genes—and move them into another organism. In 1972, the collaboration of Hubert Boyer and Stanley Cohen resulted in the first isolation and transfer of a gene from one organism to a single-celled bacterium where it would express the gene and manufacture a protein.

Their discoveries led to the first direct use of biotechnology—the production of synthetic insulin to treat people with diabetes—and the start of what is often called modern biotechnology. Plants were first transformed through genetic engineering in the late 1970s. Mary-Dell Chilton and colleagues used a common soil-dwelling bacterium, Agrobacterium tumefaciens, that attaches itself to plants and transfers some of its DNA into the plant. Chilton and her colleagues added a gene to this bacterium, which in turn transferred the gene into a plant where it became part of the plant's DNA. This bacterium is still commonly used in genetic engineering along with another technique that uses a high-velocity mechanism to inject DNA into plant cells. The result from either technique is the same—the plant cells take up the gene and begin to express it as their own.

Benefits and Risks of Biotechnology

Plants developed through genetic engineering were first grown on 1.7 million hectares in 1996 in the United States, but by 2002 they were grown on 58.7 million hectares in 16 countries. By far the major use of the present plants is to manage pests—weeds, insects and diseases. Weed management with genetically engineered plants is accomplished because the plants have a modified enzyme (a protein) that allows them to survive an application of a specific herbicide that normally acts on that enzyme. Growers can plant the herbicide-tolerant seeds, allow the plants to emerge along with any weeds in the field and then treat the field with an herbicide.

The result is that the weeds, but not the crops, die. The advantage to growers is that they spend less time on weed management, have enhanced weed control, use safer herbicides, and in many cases use less herbicides. Additionally, this technology allows growers to use soil conservation practices such as reduced or no-tillage, thus helping to retain soil structure and moisture and reduce erosion. Herbicide tolerant crops (soybean, canola, cotton and maize) were grown on 48.6 million hectares in 2002. Insect-resistant crops developed through genetic engineering utilize the common soil bacterium, Bacillus thuringiensis (Bt), which has been commercially used for more than 50 years, as an insecticide spray.

Although safe to humans and the environment, when a susceptible insect ingests Bt, the Bt protein binds to specific molecular receptors in the gut and creates pores causing the insect to starve to death. Insecticidal products containing Bt were first commercialized in France in the late 1930s, but even in 1999 the total sales of Bt products constituted less than 2 per cent of the total value of all insecticides. Bt, which had limited use as a foliar insecticide, became a major insecticide only when genes that produce Bt toxins were engineered into major crops. The Bt crops available at present are maize and cotton. These were grown on a total of 14.5 million hectares in 2002. Virus-resistant crops were created by inserting a non-infective part of a plant virus into a plant, essentially "vaccinating" the plant to protect it from the virus.

This technique is called "pathogen-derived resistance." Squash and papaya have been engineered to resist infection by some common viruses and are approved for sale in the United States. There are fewer than 1 million hectares of these crops. The bioengineered plants available at present provide growers with better tools to manage pest problems. As with any technology, there are risks and benefits to currently available genetically engineered plants, but the present body of information indicates their use has enhanced pest management, substantially reduced the amounts of pesticides used in some crops, enabled growers to use safer pesticides, and contributed to enhanced safety for humans and the environment.

The regulatory process for managing these plants and their effects on the environment and humans has evolved with the technology and the scientific community's knowledge of these tools. Many of the more controversial issues surrounding genetic engineering of plants—such as pesticide resistance, gene flow and intellectual property issues—are not unique to this new technology but pertain to all types of agriculture. Some species of insects have developed resistance to sprays of Bt, indicating the potential for some species to become resistant to Bt plants.

However, despite Bt plants being grown on more than 62 million hectares worldwide from 1996 to 2002, there have been no documented cases of resistance development. The reasons for this lack of resistance appear to involve not only biological factors of the insects and the Bt plant, but also the fact that the regulatory agency (the Environmental Protection Agency) in the United States requires a resistance management plan for growing Bt plants. No other insecticide has such strict regulations. Still, growers, companies and federal regulatory agencies must be vigilant about resistance developing for biotech crops used to manage insects, weeds and viruses as they also must with non-biotech pest management tactics. It will be important to consider the accrued environmental and health benefits of these biotech crops prior to the development of any resistance and how resistance can be managed if and where it occurs. In addition to pesticide resistance, gene flow from biotech to non-biotech crops may also be a concern. However, the risk of gene flow varies with each crop and each gene. Pollen flow in soybeans is very limited so the risk of a biotech soybean crop crossing with a non-biotech soybean is minimal, but this may be different for another crop. Likewise, if the gene in the biotech crop that provided a pest management trait, such as insect resistance, moved into a non-biotech plant, such as a weed, any selective advantage of the insect-protected weed in the ecosystem should be assessed. These same questions should also be answered with non-biotech crops, but these have not received the same level of attention as biotech crops because of the latter's higher profile.

Horizon of Biotechnology

In the future, the potential uses of plant biotechnology are far more wide-ranging than the pest-management biotech crops of today. Plants are being developed that serve as production "factories" for medically important drugs, sources of alternative energy, tools for cleaning toxic waste sites, and biomaterials including dyes, inks, detergents, adhesives, lubricants, plastics and the like. Consumers may see these products as more directly enhancing their quality of life than the pest-management biotech crops of today. Perhaps an even more dramatic advantage to consumers will be seen when plants are genetically engineered to have enhanced health benefits such as disease-fighting chemicals or increased amounts of essential vitamins and minerals. A healthy and well-informed discussion of the risks and benefits involved in agricultural biotechnology is needed to ensure a proper role for this new technology in our future food and health systems. No one should believe that any technology, including biotechnology, will completely solve the world's agricultural problems. Many people familiar with biotechnology, however, believe it to be an important component of the solution.

GENE TRANSFER METHODS IN PLANTS

To achieve genetic transformation in plants, we need the construction of a vector (genetic vehicle) which transports the genes of interest, flanked by the necessary controlling sequences *i.e.* promoter and terminator, and deliver the genes into the host plant. The two kinds of gene transfer methods in plants are:

VECTOR-MEDIATED OR INDIRECT GENE TRANSFER

Among the various vectors used in plant transformation, the Ti plasmid of Agrobacterium tumefaciens has been widely used. This bacteria is known as "natural genetic engineer" of plants because these bacteria have natural ability to transfer T-DNA of their plasmids into plant genome upon infection of cells at the wound site and cause an unorganized growth of a cell mass known as crown gall. Ti plasmids are used as gene vectors for delivering useful foreign genes into target plant cells and tissues. The foreign gene is cloned in the T-DNA region of Ti-plasmid in place of unwanted sequences.

To transform plants, leaf discs (in case of dicots) or embryogenic callus (in case of monocots) are collected and infected with Agrobacterium carrying recombinant disarmed Ti-plasmid vector. The infected tissue is then cultured (co-cultivation) on shoot regeneration medium for 2-3 days during which time the transfer of T-DNA along with foreign genes takes place. After this, the transformed tissues (leaf discs/calli) are transferred onto selection cum plant regeneration medium supplemented with usually lethal concentration of an antibiotic to selectively eliminate non-transformed tissues. After 3-5 weeks, the regenerated shoots are transferred to root-inducing medium, and after another 3-4 weeks, complete plants are transferred to soil following the hardening (acclimatization) of regenerated plants. The molecular techniques like PCR and southern hybridization are used to detect the presence of foreign genes in the transgenic plants.

VECTORLESS OR DIRECT GENE TRANSFER

In the direct gene transfer methods, the foreign gene of interest is delivered into the host plant cell without the help of a vector. The methods used for direct gene transfer in plants are:

Chemical mediated gene transfer *e.g.* chemicals like polyethylene glycol (PEG) and dextran sulphate induce DNA uptake into plant protoplasts. Calcium phosphate is also used to transfer DNA into cultured cells.

Microinjection where the DNA is directly injected into plant protoplasts or cells (specifically into the nucleus or cytoplasm) using fine tipped (0.5 - 1.0 micrometer diameter) glass needle or micropipette. This method of gene transfer is used to introduce DNA into large cells such as oocytes, eggs, and the cells of early embryo.

Electroporation involves a pulse of high voltage applied to protoplasts/cells/tissues to make transient (temporary) pores in the plasma membrane which facilitates the uptake of foreign DNA.

The cells are placed in a solution containing DNA and subjected to electrical shocks to cause holes in the membranes. The foreign DNA fragments enter through the holes into the cytoplasm and then to nucleus.

Particle gun/Particle bombardment - In this method, the foreign DNA containing the genes to be transferred is coated onto the surface of minute gold or tungsten particles (1-3 micrometers) and bombarded onto the target tissue or cells using a particle gun (also called as gene gun/shot gun/microprojectile gun).The microprojectile bombardment method was initially named as biolistics by its inventor Sanford. Two types of plant tissue are commonly used for particle bombardment- Primary explants and the proliferating embryonic tissues.

Transformation - This method is used for introducing foreign DNA into bacterial cells *e.g*. E. Coli. The transformation frequency (the fraction of cell population that can be transferred) is very good in this method. *E.g*. the uptake of plasmid DNA by E. coli is carried out in ice cold $CaCl_2$ (0-50C) followed by heat shock treatment at 37-450C for about 90 sec. The transformation efficiency refers to the number of transformants per microgram of added DNA. The $CaCl_2$ breaks the cell wall at certain regions and binds the DNA to the cell surface. Conjuction - It is a natural microbial recombination process and is used as a method for gene transfer. In conjuction, two live bacteria come together and the single stranded DNA is transferred via cytoplasmic bridges from the donor bacteria to the recipient bacteria.

Liposome mediated gene transfer or Lipofection - Liposomes are circular lipid molecules with an aqueous interior that can carry nucleic acids. Liposomes encapsulate the DNA fragments and then adher to the cell membranes and fuse with them to transfer DNA fragments. Thus, the DNA enters the cell and then to the nucleus. Lipofection is a very efficient technique used to transfer genes in bacterial, animal and plant cells.

SELECTION OF TRANSFORMED CELLS FROM UNTRANSFORMED CELLS

The selection of transformed plant cells from untransformed cells is an important step in the plant genetic engineering. For this, a marker gene (*e.g*. for antibiotic resistance) is introduced into the plant along with the transgene followed by the selection of an appropriate selection medium (containing the antibiotic). The segregation and stability of the transgene integration and expression in the subsequent generations can be studied by genetic and molecular analyses.

TRANSGENIC PLANTS WITH BENEFICIAL TRAITS

During the last decades, a tremendous progress has been made in the development of transgenic plants using the various techniques of genetic engineering. The plants, in which a functional foreign gene has been incorporated by any biotechnological methods that generally are not present in the plant, are called transgenic plants. As per estimates recorded in 2002, transgenic crops are cultivated world-wide on about 148 million acres (587 million hectares) land by about 5.5 million farmers. Transgenic plants have many beneficial traits like insect resistance, herbicide tolerance, delayed fruit ripening, improved oil quality, weed control etc.

STRESS TOLERANCE

Biotechnology strategies are being developed to overcome problems caused due to biotic stresses (viral, bacterial infections, pests and weeds) and abiotic stresses (physical actors such as temperature, humidity, salinity etc).

Abiotic Stress Tolerance

The plants show their abiotic stress response reactions by the production of stress related osmolytes like sugars (*e.g.* trehalose and fructans), sugar alcohols (*e.g.* mannitol), amino acids (*e.g.* proline, glycine, betaine) and certain proteins (*e.g.* antifreeze proteins).

Transgenic plants have been produced which over express the genes for one or more of the above mentioned compounds. Such plants show increased tolerance to environmental stresses. Resistance to abiotic stresses includes stress induced by herbicides, temperature (heat, chilling, freezing), drought, salinity, ozone and intense light. These environmental stresses result in the destruction, deterioration of crop plants which leads to low crop productivity. Several strategies have been used and developed to build ressitance in the plants against these stresses.

HERBICIDE TOLERANCE

Weeds are unwanted plants which decrease the crop yields and by competing with crop plants for light, water and nutrients. Several biotechnological strategies for weed control are being used *e.g.* the over-production of herbicide target enzyme (usually in the chloroplast) in the plant which makes the plant insensitive to the herbicide. This is done by the introduction of a modified gene that encodes for a resistant form of the enzyme targeted by the herbicide in weeds and crop plants. Roundup Ready crop plants tolerant to herbicide-Roundup, is already being used commercially.

The biological manipulations using genetic engineering to develop herbicide resistant plants are:

- Over-expression of the target protein by integrating multiple copies of the gene or by using a strong promoter.,
- Enhancing the plant detoxification system which helps in reducing the effect of herbicide.,
- Detoxifying the herbicide by using a foreign gene., and
- Modification of the target protein by mutation.

Some of the examples are:

Glyphosate resistance - Glyphosate is a glycine derivative and is a herbicide which is found to be effective against the 76 of the world's worst 78 weeds. It kills the plant by being the competitive inhibitor of the enzyme 5-enoyl-pyruvylshikimate 3- phosphate synthase (EPSPS) in the shikimic acid pathway. Due to it's structural similarity with the substrate phosphoenol pyruvate, glyphosate binds more tightly with EPSPS and thus blocks the shikimic acid pathway.

Certain strategies were used to provide glyphosate resistance to plants:

- It was found that EPSPS gene was overexpressed in Petunia due to gene amplification. EPSPS gene was isolated from Petunia and introduced in to the other plants. These plants could tolerate glyphosate at a dose of 2- 4 times higher than that required to kill wild type plants.

- *By using mutant EPSPS genes*: A single base substitution from C to T resulted in the change of an amino acid from proline to serine in EPSPS. The modified enzyme cannot bind to glyphosate and thus provides resistance.
- The detoxification of glyphosate by introducing the gene (isolated from soil organism- Ochrobactrum anthropi) encoding for glyphosate oxidase into crop plants. The enzyme glyphosate oxidase converts glyphosate to glyoxylate and amino methylphosponic acid. The transgenic plants exhibited very good glyphosate ressitance in the field.

Another Example is of Phosphinothricin Resistance

Phosphinothricin is a broad spectrum herbicide and is effective against broad-leafed weeds. It acts as a competitive inhibitor of the enzyme glutamine synthase which results in the inhibition of the enzyme glutamine synthase and accumulation of ammonia and finally the death of the plant. The disturbace in the glutamine synthesis also inhibits the photosynthetic activity. The enzyme phosphinothricin acetyl transferase (which was first observed in Streptomyces sp in natural detoxifying mechanism against phosphinothricin) acetylates phosphinothricin, and thus inactivates the herbicide. The gene encoding for phosphinothricin acetyl transferase (bar gene) was introduced in transgenic maize and oil seed rape to provide resistance against phosphinothricin.

OTHER ABIOTIC STRESSES

The abiotic stresses due to temperature, drought, and salinity are collectively also known as water deficit stresses. The plants produce osmolytes or osmoprotectants to overcome the osmotic stress. The attempts are on to use genetic engineering strategies to increase the production of osmoprotectants in the plants.

The biosynthetic pathways for the production of many osmoprotectants have been established and genes coding the key enzymes have been isolated. *E.g.* Glycine betaine is a cellular osmolyte which is produced by the participation of a number of key enzymes like choline dehydrogenase, choline monooxygenase etc. The choline oxidase gene from Arthrobacter sp. was used to produce transgenic rice with high levels of glycine betaine giving tolerance against water deficit stress.

Scientists also developed cold-tolerant genes (around 20) in Arabidopsis when this plant was gradually exposed to slowly declining temperature. By introducing the coordinating gene (it encodes a protein which acts as transcription factor for regulating the expression of cold tolerant genes), expression of cold tolerant genes was triggered giving protection to the plants against the cold temperatures.

INSECT RESISTANCE

A variety of insects, mites and nematodes significantly reduce the yield and quality of the crop plants. The conventional method is to use synthetic pesticides, which also have severe effects on human health and environment. The transgenic technology uses an innovative and eco-friendly method to improve pest control management.About 40 genes obtained from microorganisms of higher plants and animals have been used to provide insect resistance in crop plants

The first genes available for genetic engineering of crop plants for pest resistance were Cry genes (popularly known as Bt genes) from a bacterium Bacillus thuringiensis. These are specific to particular group of insect pests, and are not harmful to other useful insects like butter flies and silk worms. Transgenic crops with Bt genes (*e.g.* cotton, rice, maize, potato, tomato, brinjal, cauliflower, cabbage, etc.) have been developed. This has proved to be an effective way of controlling the insect pests and has reduced the pesticide use. The most notable example is Bt cotton (which contains CryIAc gene) that is resistant to a notorious insect pest Bollworm (Helicoperpa armigera).. There are certain other insect resistant genes from other microorganisms which have been used for this purpose. Isopentenyl transferase gene from Agrobacterium tumefaciens has been introduced into tobacco and tomato. The transenic plants with this transgene were found to reduce the leaf consumption by tobacco hornworm and decrease the survival of peach potato aphid.

Certain genes from higher plants were also found to result in the synthesis of products possessing insecticidal activity. One of the examples is the Cowpea trypsin inhibitor gene (CpTi) which was introduced into tobacco, potato, and oilseed rape for develping transgenic plants. Earlier it was observed that the wild species of cowpea plants growing in Africa were resistant to attack by a wide range of insects. It was observed that the insecticidal protein was a trypsin inhibitor that was capable of destroying insects belonging to the orders Lepidoptera, Orthaptera etc. Cowpea trypsin inhibitor (CpTi) has no effect on mammalian trypsin, hence it is non-toxic to mammals.

VIRUS RESISTANCE

There are several strategies for engineering plants for viral resistance, and these utilizes the genes from virus itself (*e.g.* the viral coat protein gene). The virus-derived resistance has given promising results in a number of crop plants such as tobacco, tomato, potato, alfalfa, and papaya. The induction of virus resistance is done by employing virus-encoded genes-virus coat proteins, movement proteins, transmission proteins, satellite RNa, antisense RNAs, and ribozymes. The virus coat protein-mediated approach is the most successful one to provide virus resistance to plants. It was in 1986, transgenic tobacco plants expressing tobacco mosaic virus (TMV) coat protein gene were first developed. These plants exhibited high levels of resistance to TMV.

The transgenic plant providing coat protein-mediated resistance to virus are rice, potato, peanut, sugar beet, alfalfa etc. The viruses that have been used include alfalfa mosaic virus (AIMV), cucumber mosaic virus (CMV), potato virus X (PVX) , potato virus Y (PVY) etc.

RESISTANCE AGAINST FUNGAL AND BACTERIAL INFECTIONS

As a defence strategy against the invading pathogens (fungi and bacteria) the plants accumulate low molecular weight proteins which are collectively known as pathogenesis-related (PR) proteins.

Several transgenic crop plants with increased resistance to fungal pathogens are being raised with genes coding for the different compounds. One of the examples is the Glucanase enzyme that degrades the cell wall of many fungi. The most widely used glucanase is beta-1,4-glucanase. The gene encoding for beta-1,4 glucanase has been isolated from barley, introduced, and expressed in transgenic tobacco plants. This gene provided good protection against soil-borne fungal pathogen Rhizoctonia solani.

Lysozyme degrades chitin and peptidoglycan of cell wall, and in this way fungal infection can be reduced. Transgenic potato plants with lysozyme gene providing resistance to Eswinia carotovora have been developed.

DELAYED FRUIT RIPENING

The gas hormone, ethylene regulates the ripening of fruits, therefore, ripening can be slowed down by blocking or reducing ethylene production. This can be achieved by introducing ethylene forming gene(s) in a way that will suppress its own expression in the crop plant. Such fruits ripen very slowly (however, they can be ripen by ethylene application) and this helps in exporting the fruits to longer distances without spoilage due to longer-shelf life.

The most common example is the 'Flavr Savr' transgenic tomatoes, which were commercialized in U.S.A in 1994. The main strategy used was the antisense RNA approach. In the normal tomato plant, the PG gene (for the enzyme polygalacturonase) encodes a normal mRNA that produces the enzyme polygalacturonase which is involved in the fruit ripening.

The complimentary DNA of PG encodes for antisense mRNA, which is complimentary to normal (sense) mRNA. The hybridization between the sense and antisnse mRNAs renders the sense mRNA ineffective. Consequently, polygalacturonase is not produced causing delay in the fruit ripening. Similarly strategies have been developed to block the ethylene biosynthesis thereby reducing the fruit ripening. *E.g.* transgenic plants with antisense gene of ACC oxidase (an enzyme involved in the biosynthetic process of ethylene) have been developed. In these plants, production of ethylene was reduced by about 97% with a significant delay in the fruit ripening.

The bacterial gene encoding ACC deaminase (an enzyme that acts on ACC and removes amino group) has been transferred and expressed in tomato plants which showed 90% inhibition in the ethylene biosynthesis.

BIOTECHNOLOGY IN PLANT CELL

Clonal propagation refers to the process of asexual reproduction by multiplication of genetically identical copies of individual plants. The vegetative propagation of plants is labour-intensive, low in productivity and seasonal. The tissue culture methods of plant propagation, known as 'micropropagation' utilizes the culture of apical shoots, axillary buds and meristems on suitable nutrient medium.The regeneration of plantlets in cultured tissue was described by Murashige in 1974. Fossard gave a detailed account of stages of micropropagation.

The micropropagation is rapid and has been adopted for commercialization of important plants such as banana, apple, pears, strawberry, cardamom, many ornamentals (*e.g.* Orchids) and other plants.The micropropagation techniques are preferred over the conventional asexual propagation methods because of the following reasons: (a) In the micropropagation method, only a small amount of tissue is required to regenerate millions of clonal plants in a year., (b) micropropagation is also used as a method to develop resistance in many species., (c) in vitro stock can be quickly proliferated as it is season independent,. (d) long term storage of valuable germplasm possible.

The steps in micropropagation method are:

- *Initiation of culture*: From an explant like shoot tip on a suitable nutrient medium,
- Multiple shoots formation from the cultured explant,
- Rrooting of in vitro developed shoots and,
- Transplantation - transplantation to the field following acclimatization.

The factors that affect micropropagation are:

- Genotype and the physiological status of the plant *e.g.* plants with vigorous germination are more suitable for micropropagation.,
- The culture medium and the culture environment like light, temperature etc. For example an illumination of 16 hours a day and 8 hours night is satisfactory for shoot proliferation and a temperature of 25°C is optimal for the growth.

The benefits of micropropagation this method are:

- Rapid multiplication of superior clones can be carried out through out the year, irrespective of seasonal variations.
- Multiplication of disease free plants *e.g.* virus free plants of sweet potato (Ipomea batatus), cassava (Manihot esculenta)

- Multiplication of sexually derived sterile hybrids
- It is a cost effective process as it requires minimum growing space.

SOMACLONAL VARIATION

The genetic variations found in the in vitro cultured cells are collectively referred to as somaclonal variation and the plants derived from such cells are called as 'somaclones'. It has been observed that the long-term callus and cell suspension culture and plants regenerated from such cultures are often associated with chromosomal variations. It is this property of cultured cells that finds potential application in the crop improvement and in the production of mutants and variants (*e.g.* disease resistance in potato).

Larkin and Scowcroft working at the division of Plant Industry, C.S.I.R.O., Australia gave the term 'somaclones' for plant variants obtained from tissue cultures of somatic tissues. Similarly, if the tissue from which the variants have been obtained is having gametophytic origin such as pollen or egg cell, it is known as 'gametoclonal' variation.

They explained that it may be due to:

- Reflection of heterogeneity between the cells and explant tissue,
- A simple representation of spontaneous mutation rate, and
- Activation by culture environment of transposition of genetic materials.

Shepard *et al.* also contributed by screening about 100 somaclones produced from leaf protoplasts of Russet Burbank. They found that there was a significant amount of stable variation in compactness of growth habit, maturity, date, tuber uniformity, tuber skin colour and photoperiodic requirements. Somaclonal Variations has been used in plant breeding programmes where the genetic variations with desired or improved characters are introduced into the plants and new varieties are created that can exhibit disease resistance, improved quality and yield in plants like cereals, legumes, oil seeds tuber crops etc. Somaclonal variation is applicable for seed

APPLICATIONS OF SOMACLONAL VARIATIONS

- Methodology of introducing somaclonal variations is simpler and easier as compared to recombinant DNA technology.
- Development and production of plants with disease resistance *e.g.* rice, wheat, apple, tomato etc.
- Develop biochemical mutants with abiotic stress resistance *e.g.* aluminium tolerance in carrot, salt tolerance in tobacco and maize.
- Development of somaclonal variants with herbicide resistance *e.g.* tobacco resistant to sulfonylurea
- Development of seeds with improved quality *e.g.* a new variety of Lathyrus sativa seeds (Lathyrus Bio L 212) with low content of neurotoxin.

- Bio-13 – A somaclonal variant of Citronella java (with 37% more oil and 39% more citronellon), a medicinal plant has been released as Bio-13 for commercial cultivation by Central Institute for Medicinal and Aromatic Plants.
- Supertomatoes- Heinz Co. and DNA plant Technology Laboratories (USA) developed Supertomatoes with high solid component by screening somaclones which helped in reducing the shipping and processing costs.

Production of Virus Free Plants

The viral diseases in plants transfer easily and lower the quality and yield of the plants. It is very difficult to treat and cure the virus infected plants therefore te plant breeders are always interested in developing and growing virus free plants.

In some crops like ornamental plants, it has become possible to produce virus free plants through tissue culture at the commercial level.

This is done by regenerating plants from cultured tissues derived from:

- Virus free plants,
- Meristems which are generally free of infection - In the elimination of the virus, the size of the meristem used in cultures play a very critical role because most of the viruses exist by establishing a gradient in plant tissues. The regeneration of virus-free plants through cultures is inversely proportional to the size of the meristem used.,
- Meristems treated with heat shock (34-36°C) to inactivate the virus,
- Callus, which is usually virus free like meristems.
- Chemical treatment of the media- attempts have been made to eradicate the viruses from infected plants by treating the culture medium with chemicals *e.g.* addition of cytokinins suppressed the multiplication of certain viruses.

Among the culture techniques, meristem-tip culture is the most reliable method for virus and other pathogen elimination. Viruses have been eliminated from a number of economically important plant species which has resulted in a significant increase in the yield and production *e.g.* potato virus X from potato, mosaic virus from cassava etc. These virus free plants are not disease resistant so there is a need to maintain stock plants to multiply virus free plants whenever required.

Production of Synthetic Seeds

In synthetic seeds, the somatic embryos are encapsulated in a suitable matrix (*e.g.* sodium alginate), along with substances like mycorrhizae, insecticides, fungicides and herbicides. These artificial seeds can be utilized for the rapid and mass propagation of desired plant species as well as hybrid varieties.

The major benefits of synthetic seeds are:

- They can be stored up to a year with out loss of viability
- Easy to handle and useful as units of delivery
- Can be directly sown in the soil like natural seeds and do not need acclimatization in green house.

Mutant Selection

An important use of cell cultures is in mutant selection in relation to crop improvement. The frequency of mutations can be increased several fold through mutagenic treatments and millions of cells can be screened. A large number of reports are available where mutants have been selected at cellular level. The cells are often selected directly by adding the toxic substance against which resistance is sought in the mutant cells. Using this method, cell lines resistant to amino acid analogues, antibiotics, herbicides, fungal toxins etc have actually been isolated.

Production of Secondary Metabolites

The most important chemicals produced using cell culture are secondary metabolites, which are defined as' those cell constituents which are not essential for survival'. These secondary metabolites include alkaloids, glycosides (steroids and phenolics), terpenoids, latex, tannins etc. It has been observed that as the cells undergo morphological differentiation and maturation during plant growth, some of the cells specialize to produce secondary metabolites. The in vitro production of secondary metabolites is much higher from differentiated tissues when compared to non-differentiated tissues. The cell cultures contribute in several ways to the production of natural products.

These are:

- A new route of synthesis to establish products *e.g.* codeine, quinine, pyrethroids,
- A route of synthesis to a novel product from plants difficult to grow or establish *e.g.* thebain from Papaver bracteatum,
- A source of novel chemicals in their own right *e.g.* rutacultin from culture of Ruta,
- As biotransformation systems either on their own or as part of a larger chemical process *e.g.* digoxin synthesis.

The Advantages of In Vitro Production of Secondary Metabolites

- The cell cultures and cell growth are easily controlled in order to facilitate improved product formation.
- The recovery of the product is easy.
- As the cell culture systems are independent of environmental factors, seasonal variations, pest and microbial diseases, geographical

location constraints, it is easy to increase the production of the required metabolite.

- Mutant cell lines can be developed for the production of novel and commercially useful compounds.
- Compounds are produced under controlled conditions as per the market demands.
- The production time is less and cost effective due to minimal labour involved.

APPLICATIONS OF SECONDARY METABOLITES

Many of these secondary products especially various alkaloids are of immense use in medicine.

The yield of these chemicals in cell culture, is though generally lower than in whole plants, it is substantially increased by manipulating physiological and biochemical conditions.

Shikonine is a dye produced by the cells Lithospermum erythrorhizon on a commercial scale. Besides this there are a number of secondary metabolite products that are being widely used for various purposes. Vincristine is used as anticancer agent, digoxin controls cardiovascular disorders, pyrithrins is an insecticide etc. The production of specialty chemicals by plants has become a multibillion industry. Please refer to the table for some secondary metabolites and their uses.

Table. Showing Plant Species and Secondary Metabolites Obtained from them Using Tissue Culture Techniques

Product	Plant Source	Uses
Artemisin	Artemisia spp.	Antimalarial
Azadirachtin	Azadirachta indica	Insecticidal
Berberine	Coptis japonica	Antibacterial, anti inflammatory
Capsaicin	Capsicum annum	Cures Rheumatic pain
Codeine	Papaver spp.	Analgesic
Camptothecin	Campatotheca accuminata	Anticancer
Cephalotaxine	Cephalotaxus harringtonia	Antitumour
Digoxin	Digitalis lanata	Cardiac tonic
Pyrethrin	Chrysanthemum cinerariaefolium	Insecticide (for grain storage)
Morphine	Papaver somniferum	Analgesic, sedative
Quinine	Cinchona officinalis	Antimalarial
Taxol	Taxus spp.	Anticarcinogenic
Vincristine	Cathranthus roseus	Anticarcinogenic
Scopolamine	Datura stramonium	Antihypertensive

Production of Somatic Hybrids and Cybrids

The Somatic cell hybridization/parasexual hybridization or Protoplast fusion offers an alternative method for obtaining distant hybrids with desirable traits significantly between species or genera, which can not be made to cross by conventional method of sexual hybridization.

SOMATIC HYBRIDIZATION

Somatic hybridization broadly involves in vitro fusion of isolated protoplasts to form a hybrid cell and its subsequent development to form a hybrid plant.

The process involves:

- Fusion of protoplasts,
- Selection of hybrid cells,
- Identification of hybrid plants.

During the last two decades, a variety of treatments have been used to bring about the fusion of plant protoplasts. Protoplast fusion can be achieved by spontaneous, mechanical, or induced fusion methods..

These treatments include the use of fusogens like $NaNO_3$, high pH with high Ca^{2++} ion concentration, use of polyethylene glycol (PEG), and electrofusion. These inducing agents used in protoplast fusion are called 'fusogen'.

PEG treatment is the most widely used method for protoplast fusion as it has certain advantages over others.

These are:

- It results in a reproducible high-frequency of heterokaryon formation.,
- The PEG fusion is non specific and therefore can be used for a wide range of plants.,
- It has low toxicity to the cell and
- The formation of binucleate heterokaryons is low.

MECHANISM OF FUSION

The fusion of protoplasts takes place in three phases- agglutination, plasma membrane fusion and formation of heterokaryons. When the two protoplasts come in close contact with each other, they adhere to each other. This agglutination can be induced by PEG, high pH and high Ca^{2+}. The protoplast membranes get fused at localized sites at the point of adhesion. This leads to the formation of cytoplasmic bridges between protoplasts.

High pH and high Ca^{2+} ions neutralize the surface charges on the protoplasts which allows closer contact and membrane fusion between agglutinated protoplasts. The fused protoplasts become round as a result of cytoplasmic bridges which leads to the formation of spherical homokaryon or heterokaryon.

SELECTION OF HYBRID CELLS

The methods used for the selection of hybrid cells are biochemical, visual and cytometric methods using fluorescent dyes. The biochemical methods for selection of hybrid cells are based on the use of biochemical compounds in the medium. The drug sensitivity method is useful for the selection hybrids of two plants species, if one of them is sensitive to a drug. Another method, auxotrophic mutant selection method involves the auxotrophs which are mutants that cannot grow on a minimal medium. Therefore specific compounds are added in the medium.

The selection of auxotropic mutants is possible only if the hybrid cells can grow on a minimal medium. The visual method involves the identification of heterokaryons under the light microscope. In some of the somatic hybridizations, the chloroplast deficient protoplast of one plant species is fused with the green protoplast of another plant species. The heterokaryons obtained are bigger and green in colour while the parental protoplasts are either small or colourless. The cytometric method uses flow cytometry and flourescent-activated cell sorting techniques for the analysis of plant protoplasts.

APPLICATIONS OF SOMATIC HYBRIDIZATION

- *Creation of hybrids with disease resistance*: Many disease resistance genes (*e.g.* tobacco mosaic virus, potato virus X, club rot disease) could be successfully transferred from one species to another. *E.g* resistance has been introduced in tomato against diseases such as TMV, spotted wilt virus and insect pests.
- *Environmental tolerance*: Using somatic hybridization the genes conferring tolerance for cold, frost and salt were introduced in *e.g.* in tomato.
- *Cytoplasmic male sterility*: Using cybridization method, it was possible to transfer cytoplasmic male sterility.
- *Quality characters*: Somatic hybrids with selective characteristics have been developed *e.g.* the production of high nicotine content.

6

Agricultural Biotechnology Sector in India

Agricultural biotechnology sector in India, particularly its plant biotechnology segment, is uniquely poised for a major process of transformation. As the country's agricultural biotechnology sector made strides in the 1990s in r-DNA, transgenics and molecular marker assisted plant breeding process, the Government of India responded with a matching policy support and regulatory framework that was designed to render the path of progress in R&D, sustainable and bio-safe. To a large extent, developments in the policy front have been induced by a vibrant non-governmental sector that intensely intervened on the sensitivities of modern biotechnology. This chapter discusses analysis of the issues and proceeds to further consider the path of progress achieved by agricultural biotechnology companies in India in the area of commercialization of biotechnology products. The special focus of this chapter is on plant biotechnology. The biotechnology sector in India in 2002-03 is estimated to have attracted private investments to the tune of US$ 140 million. Infrastructure and R&D account for nearly 80 per cent of these investments.

During the period from 1986 to 2002 it is estimated that the Government of India invested an amount of US$ 275 million in the biotechnology sector. The Government of India has also extended venture capital support for biotechnology start-ups through the Technology Development Board and the New Millennium Indian Technology Industry Leadership Initiative (NMITLI). Meanwhile the States of Andhra Pradesh, Karnataka, Maharashtra, Gujarat, Kerala and Tamil Nadu have drawn up ambitious sub-national biotechnology policies, which promise fiscal and infrastructure support measures to prospective entrepreneurs. Some State Governments have launched their own venture funds for supporting novel start-up ventures.

To catalyse the process, the Government of India accorded priority to IPR protection. The Indian Patents Act of 1970 was amended to provide for product patents for agro-chemicals, drugs and pharmaceuticals and micro-organisms over a transitory period. A product patent regime would be in place in India by January 1, 2005. The new patent law extends protection term for inventions from 14 to 20 years. To provide for protection of IPRs in respect of new plant varieties, Parliament has enacted a *sui-generis* legislation in the shape of the Plant Varieties Protection and Farmers Rights (PVFR) Act in 2001. These developments have created favourable legal conditions for international partnerships in biotechnology R&D. By prioritizing FDI investments in biotechnology, the Government of India has sent appropriate signals to international investors. Further the fact that these changes have been in consonance with the WTO-TRIPs has lent greater creditability to these changes. Meanwhile, the Government has also brought in regulations over agricultural biotechnology. Apart from the biosafety regulations that form part of the Environment (Protection) Act 1986, the Indian Parliament has legislated the National Biodiversity Act in the year 2002, to provide for regulatory controls over access to biological resources in India.

AGRICULTURE IN INDIA

India's record of progress in agriculture over the past four decades has been quite impressive. The agriculture sector has been successful in keeping pace with rising demand for food. The contribution of increased land area under agricultural production has declined over time and increases in production in the past two decades have been almost entirely due to increased productivity.

Contribution of agricultural growth to overall progress has been widespread. Increased productivity has helped to feed the poor, enhanced farm income and provided opportunities for both direct and indirect employment. The success of India's agriculture is attributed to a series of steps that led to availability of farm technologies which brought about dramatic increases in productivity in 70s and 80s often described as the Green Revolution era. The major sources of agricultural growth during this period were the spread of modern crop varieties, intensification of input use and investments leading to expansion in the irrigated area. In areas where 'Green Revolution' technologies had major impact, growth has now slowed.

New technologies are needed to push out yield frontiers, utilize inputs more efficiently and diversify to more sustainable and higher value cropping patterns. At the same time there is urgency to better exploit potential of rainfed and other less endowed areas if we are to meet targets of agricultural growth and poverty alleviation. Given the wide range of agroecological setting and producers, Indian agriculture is faced with a great diversity of needs, opportunities and prospects.

Future growth needs to be more rapid, more widely distributed and better targeted. These challenges have profound implications for the way farmers' problems are conceived, researched and transferred to the farmers. On the one hand agricultural research will increasingly be required to address location specific problems facing the communities on the other the systems will have to position themselves in an increasingly competitive environment to generate and adopt cutting edge technologies to bear upon the solutions facing a vast majority of resource poor farmers. In the past agriculture has played and will continue to play a dominant role in the growth of Indian economy in the foreseeable future. It represents the largest sector producing around 28 percent of the GDP, is the largest employer providing more than 60 percent of the jobs and is the prime arbiter of living standards for seventy percent of India's population living in the rural areas.

These factors together with a strong determination to achieve self-sufficiency in food grains production have ensured a high priority for agriculture sector in the successive development plans of the country.

An important facet of progress in agriculture is its success in eradication of its critical dependence on imported foodgrains. In the 1950's nearly 5 percent of the total foodgrains available in the country were imported. This dependence worsened during the 1960's when two severe drought years led to a sharp increase in import of foodgrains. During 1966 India had to import more than 10 million tonnes of foodgrains as against a domestic production of 72 million tonnes. In the following year again, nearly twelve million tonnes had to be imported. On the average well over seven percent of the total availability of foodgrains during the 1960s had to be imported.

Indian agriculture has progressed a long way from an era of frequent droughts and vulnerability to food shortages to becoming a significant exporter of agricultural commodities. This has been possible due to persistent efforts at harnessing the potential of land and water resources for agricultural purposes. Indian agriculture, which grew at the rate of about 1 percent per annum during the fifty years before independence, has grown at the rate of about 3 percent per annum in the post independence era.

AGRICULTURE – SUB-SECTORS

Indian agriculture broadly consists of four sub-sectors. Agriculture proper including all food-crops oilseeds, fiber, plantation crops, fruits and vegetables is the largest accounting for nearly 70 percent of the agriculture sector as a whole.

Table. Agriculture sub-sectors – value of output (%)

Sub-sector	1960-61	1970-71	1980-81	1993-94
Agriculture	78.0	79.3	75.3	68.5
Livestock	16.6	15.1	17.2	24.5
Forestry	4.4	4.4	5.9	3.8
Fisheries	1.0	1.2	1.6	3.2
	100.0	100.0	100.0	100.0
Total value at Current price (₹ Billion)	89	218	615	2740

The rapid growth in this sub-sector through exploitation of wastelands and fallows, spread of irrigation and adoption of production enhancing technologies was critical in transforming India from a country vulnerable to food shortages to one of exportable surplus. Although this sub-sector has made impressive progress its share in the sector as a whole has declined from 78 percent in 1960-61 to less than 70 percent by early 90s.

Correspondingly the share of livestock sector has increased considerably. The livestock industry has grown from ₹ 15 billion in early 1960s to ₹ 100 billion by 1980-81 and ₹ 672 billion by 1993-94. In nominal terms the sector grew at almost 15 percent per annum during 1980s. Milk production, which was almost stagnant for two decades ending 1970, grew by over 5 percent per annum in the 80s. Similarly, production of eggs increased at the rate of about 6.5 percent during the same period. As a result the share of livestock increased from about 17 percent till early 80s to 25 percent by 1993-94.

Though it plays relatively a minor role within the sector as a whole, fishing sub-sector activities have been on the rise. The sub-sector has grown from only ₹ 3 billion in 1970-71 to nearly ₹ 90 billion in 1993-94. The growth was particularly rapid in 70s and 80s. Value added increased at over 5 percent per annum during this period.

In real terms forestry and logging activities have been on the decline since mid seventies. As of 1993-94, the size of the industry in terms of value of output was 103 billion.

Table. Foodgrains Production

Year	Area (million ha)	Production (million tonnes)	Yield kg/ha
1950-51	97.32	50.82	522
1960-61	115.58	82.02	709
1970-71	124.32	108.42	872
1980-81	126.67	129.59	1023
1990-91	127.84	176.39	1379
1998-99	124.0	195.25	1574

Over the past three decades, the country has successfully transformed itself from a food deficit economy to one which is essentially self sufficient in availability of foodgrains and other essential commodities, albeit only at the prevailing level of effective demand. Annual aggregate foodgrains production, which averaged about 82 million tonnes in 1960-61 increased to 123.7 and 172.5 million tonnes for the trienniums ending 1980-81 and 1990-91 respectively. Current (1998-99) production level is 195 million tonnes and the country has been able to accumulate substantial, (35 million tonnes) stocks of foodgrains to cope up with any sudden difficulties arising from drought or a similar situation in any part of the country.

Increased outputs, have been achieved chiefly by adopting, since mid sixties, a strategy aimed at increasing foodgrains production by concentrating public sector efforts and resources in regions with a high potential for quick and substantial productivity gains through increased cropping intensity and average yields. These were the areas favoured by agroclimatic resource conditions and where irrigation facilities already existed or could be developed relatively rapidly. The main elements of this strategy were:

- Expansion of irrigation coverage,
- Increased provision and utilization of key inputs – mainly high yielding varieties (HYVs) of crops, mainly of wheat and rice and chemical fertilizers and plant protection chemicals,
- Expansion and improvement of institutional support services such as research and extension and
- Price policies favourable to producers of major foodgrains.

The success of this strategy was made possible by development and availability of replicable production technology packages, so called 'Green Revolution' technologies. Irrigation facilitated double cropping and widespread adoption of HYVs. The HYVs performed particularly well under irrigated conditions, were highly responsive to fertilizers and their short duration permitted increases in cropping intensities.

Irrigation development was the cornerstone of the strategy. Undivided India was amongst the largest irrigated areas in the world. With partition nearly one-third of the irrigated area went to Pakistan. At the time of independence the net irrigated area was 20.9 million ha (gross irrigated area 22.6 million ha). Recognizing large-scale development of irrigation facilities as critical to rapid agricultural growth, the country has spent about ₹ 45,000 crores on irrigation development in the first four decades after independence. During the period 1950-51 to 1965-66 development of irrigation through government canals grew from 7.2 million ha to 9.8 million ha – a growth rate of 2.1 percent per annum. During 1970s this pace dropped slightly to 1.9 percent. In 1980s the rate of increase dropped significantly to 1.1 percent per annum.

Table. India – irrigated area

Year	Net	Gross	Net as % of net sown	Gross as % of gross cropped
	––– million ha –––		––––– % –––––	
1970-71	31.10	38.19	22.17	23.04
1980-81	38.72	49.78	27.66	28.83
1990-91	48.02	63.20	33.61	34.03
1995-96	53.51	71.51	37.63	38.33

The growth of tube-well irrigation, however, increased rapidly from 4.5 million ha in 1970-71 to 9.5 million ha in 1980-81 and then to 14.3 million ha by 1990-91. The net irrigated area increased from 31 million ha in 1970-71 to 53.5 million ha in 1995-96 which corresponds to 22 percent of the net sown area in 1970-71 and 37.63 percent in 1995-96 *(Tables 3, 4)*. With improvements in irrigation efficiency the gross irrigated areas has increased to 71.51 million ha. The percentage of gross cropped area service by irrigation increased from 18.3 percent in 1960-61 to 23.0 percent in 1970-71 and to over 38 percent at present.

Table. Sources of irrigation

Year	Net Irrigated Area (million ha)	Sources				
		Canals	Tanks	tube-wells	Other wells	Other Sources
		(% net irrigated area)				
1960-61	24.80	10.4	4.6	0.2	7.2	2.4
1970-71	31.10	41.28	13.22	14.34	23.88	7.29
1980-81	38.72	39.49	8.22	24.62	21.08	6.59
1990-91	48.02	36.34	6.13	29.69	21.73	6.11
1995-96	53.01	32.04	5.81	33.52	22.16	6.47

Fertilizers have constituted yet another key input in addition to expanded irrigation and spread of HYVs in achieving goals of high production and productivity. India currently occupies third position in the world, after China and USA, in terms of fertilizer production and consumption. Consumption of fertilizers has increased from 1.54 million tonnes in 1967-68, representing the pre green revolution era, to 17.31 million tonnes currently (1997-98).

The average per hectare use of fertilizers currently around 85 kg per hectares is the lowest among several Asian countries.

However, rice and wheat account for a major fraction, around 65 percent of the total fertilizer consumed in the country, with very little fertilizers going to the rainfed areas. According to some current projections, fertilizer's use will need to increase to 30-35 million tonnes to meet the foodgrains need of 2020. The demand for nutrients will stretch by almost another 15 million tonnes if requirements for horticulture, vegetables and plantation and commercial crops are included. At present domestic production of N and P fertilizers (13.42 million tonnes) falls short of consumption by over 20 percent. In addition the entire requirement of K fertilizer is imported.

Most agricultural development programmes initiated in 1960s were concentrated in regions of high potential. Thus five states, Punjab, Haryana, Uttar Pradesh, Andhra Pradesh and Tamil Nadu account for 50 percent of the country's net irrigated and 53 percent of the gross irrigated area. The combination of expanding irrigation coverage and widespread adoption of short duration HYVs led to significant increases in cropping intensities. Acreage cropped more than once per year increased from 13 million ha in 1950-51 to about 44 million ha at present. Average cropping intensity for the country as a whole rose from 115 percent in 1960-61 to 131 in 1993-94. By 1993-94 cropping intensity has risen to 187 percent in Punjab, 167 percent in Haryana and 142 percent in Uttar Pradesh.

Table. Cropping intensity

Years	Net sown area	Gross cropped area	Area cropped more than once	Cropping intensity (%)
	million ha			
1950-51	118.8	131.9	13.1	111.1
1960-61	133.2	152.8	19.6	114.3
1970-71	140.3	165.8	25.5	118.2
1980-81	140.0	172.6	32.6	123.3
1990-91	143.0	185.7	42.7	129.9
1993-94	142.0	186.4	44.4	131.2

An important consequence of the strategies adopted since sixties has been to boost production of, chiefly, two crops rice and wheat. Their share in total foodgrains production went up from 57 percent in 1970-71 to more than 75 percent in 1990-91. Production of foodgrains other than rice and wheat did not increase significantly and in the eastern region even the yield of rice did not increase. Agricultural production and income rose substantially in the north-western states of Punjab, Haryana, Western Uttar Pradesh, parts of Rajesthan, Tamil Nadu and Andhra Pradesh. By contrast productivity and output growth have been modest in eastern and central India and in deccan plateau. Progress was particularly slow in rainfed areas, which account for over 60 percent of the cropped area and where a great majority of rural poor are concentrated. An important impact of the strategies pursued in the 'Green Revolution' period has been intensification of regional disparities and imbalances in agricultural development and food availability and hence levels of food security. For the country as a whole while per capita availability of cereals has increased substantially, that of pulses has decreased significantly.

In summary an annual increase in foodgrains production of 3.22 percent during fifties was mainly because of expansion in area. Sixties recorded a low annual growth rate of 1.72 percent necessitating large-scale imports of foodgrains.

Table. Rice and Wheat Production
Production (million tonnes)

Year	Rice	Wheat	Foodgrains	R+W as % of Foodgrains
		million tonnes		
1970-71	40.80	20.86	108.4	57.4
1980-81	49.91	34.55	129.6	65.4
1990-91	72.78	53.03	176.4	75.0
1996-97	81.74	69.35	199.4	75.0

Annual growth of 2.08 percent was recorded during seventies. This decade was the turning point in India's foodgrains economy leading to self-sufficiency through significant productivity increase first in wheat and later in rice in the eighties. An annual growth of 3.5 percent in foodgrains in eighties was the hallmark of green revolution that enabled India to become self-sufficient and even a marginal exporter. The pace of growth slowed in nineties barely making or even slower than the population growth rate. This is a matter of concern.

Table. Annual compound growth rate of foodgrains production

Crop	1950-51 to 1959-60	1960-61 to 1969-70	1970-71 to 1979-80	1980-81 to 1989-90	1990-91 to 1997-98
Rice	3.28	- 8.05	1.91	4.29	1.53
Wheat	4.51	5.90	4.69	4.24	3.67
Coarse cereals	2.75	1.48	0.74	0.74	- 0.49
Total cereals	3.00	2.51	2.37	3.63	1.84
Pulses	2.72	1.35	- 0.54	2.78	0.76
Total foodgrains	3.22	1.72	2.08	3.54	1.66

Table. Per capita net availability of cereals and pulses per day (g)

Year	*Cereals*	*Pulses*	*Total*
1951	334	61	395
1961	400	69	469
1971	418	51	469
1981	417	38	455
1991	468	42	510
1998	451	33	484

HORTICULTURE

The diversity of physiographic, climate and soil characteristics enables India to grow a large variety of horticultural crops – fruits, vegetables, flowers, spices, aromatic and medicinal plants, plantation crops etc. India is the largest producer of fruits in the world and second largest producer of vegetables. The area under fruits estimated at 1.45 million ha in 1970-71 grew to 2.8 million ha in 1991-92 and then more rapidly to 5 to 6 million ha by 1994-95. This sector is likely to grow rapidly in the future both on account of internal demands and export opportunities.

ANIMAL HUSBANDRY AND FISHERIES

Animal husbandry and dairying sub-sector plays an important role in overall economy and in social development. The contribution of the sub-sector is estimated to be about 25 percent of the total value of output of agricultural sector.

Table. Production of milk, eggs and fish in India

Year	Milk production / availability		Fish	Eggs
	million tonnes/annum	g./day	(000 tonnes)	(million)
1950-51	17.0	124	752	1832
1960-61	20.0	124	1160	2881
1970-71	22.0	112	1756	6172
1980-81	31.6	128	2442	10060
1990-91	53.9	176	3836	21101
1998-99	75.0	210	5388	28400

The sector also plays a significant role in supplementing family incomes and generating employment in the rural sector particularly among the land-less, small and marginal farmers and women besides providing nutritious food. Production of suitable cross breeds and their wider adoptions has contributed to increasing country's milk production, which has now reached 75 million tonnes annually. Similarly, genetic improvement and better management practices have markedly pushed up production of poultry and eggs.

Table. Number of holdings (million)

Category of Farmers	1970-71	1980-81	1990-91
Marginal	35.7 (50.6)*	50.1 (56.4)	62.1 (59.0)
Small	13.4 (19.0)	16.1 (18.1)	20.0 (19.0)
Semi-medium	10.7 (15.2)	12.5 (14.0)	13.9 (13.2)
Medium	7.9 (11.2)	8.1 (9.1)	7.6 (7.3)
Large	2.8 (3.9)	2.2 (2.4)	1.7 (1.6)
All groups	70.5 (100)	89.0 (100)	105.3 (100)

Note:

*Figures in parenthesis represent the percentage of holdings.

Through the overall contribution of fisheries sub-sector is small, development of prolific and fast growing forms of several common fish species coupled with breakthrough in breeding under captivity, fish seed production and multi-layer fish culture has resulted in registering a very high annual growth of 11 percent in aquaculture production during the past decade.

Table. Area operated - % of total area

Category of Farmers	*1970-71*	*1980-81*	*1990-91*
Marginal	9.0	12.1	14.9
Small	11.9	14.2	17.3

Semi-medium	18.5	21.1	23.1
Medium	29.7	29.6	27.2
Large	30.9	23.0	17.4
All groups (million ha)	**162.1 (100)**	**163.8 (100)**	**165.6 (100)**

AGRICULTURAL BIOTECHNOLOGY INDUSTRY

The top ten pesticide and seed firms for 1997 and 1999. Longer-term time series data on the structure of the chemical and seed company are difficult to obtain. In one of the few published papers to provide this data, Ollinger and Fernandez-Cornejo (1998) provide Concentration Ratio data (CR4) for the United States pesticide industry over the period 1972-1989. During this period, the CR4 oscillates, moving from a high of 50% in 1973 to a low of 37% in 1982, and then rising to 48% in 1989. Although detailed data are not available for the 1990s, there is evidence that the structure of the seed and chemical industry remains dynamic.

The consolidation activity of the ten most active biotechnology firms. Mergers and acquisitions are a key component of the new business relationships that have emerged. As Kalaitzandonakes and Hayenga (2000) note, firm entry in the crop biotechnology industry peaked in the early 1980s, with production innovation continuing throughout the 1980s. Oehmke et al. (1999) also discuss the cyclical pattern in mergers in the biotechnology industry. Aggregate market figures mask the much higher concentration that exists in specific markets. For example, in 1998, Monsanto and Pioneer-HiBred (now owned wholly by DuPont) controlled 15% and 39% of the US seed corn market, respectively. These two same companies controlled approximately 24% and 17%, respectively, of the purchased soybean seed market. For the US cotton market, two companies, Delta & Pine Land and Stoneville, had 71% and 16%, respectively, of the seed market.

The determination of the relevant market concentration is not always done on the basis of output markets. The Federal Trade Commission has used innovation competition to assess the impact of mergers. Examining competition in innovation is designed to focus attention on the impact of a merger on innovative activity. Using data on field trials underway each year, Brennan Pray, and Courtmanche (2000) calculate a CR4 ratio at the R&D stage. In 1988, the four largest firms had 87% of the field trials. The CR4 ratio declined to a low of 63% in 1995, then rose steadily over the next few years to reach 79% in 1998. This concentration at the R&D stage is matched by concentration in terms of the number of patents held. The top four firms held 41% of the corn patents (up to 1996), 53% of the soybean patents (up to 1997), 77% of the tomato patents (up to June 1997) and 38% of the Bt patents (up to 1998).

Concentrated markets do not necessarily imply the presence of market power. Baumol, Panzar and Willig (1992) stress that firms will not be able to exercise market power if the market is contestable. The key requirements for market contestability are: (a) Potential entrants must not be at a cost disadvantage to existing firms, and (b) entry and exit must be costless. For entry and exit to be costless, there must be no sunk costs. Sunk costs are expenditures that cannot be recouped once they are incurred—examples include expenditures made to obtain regulatory approval, expenditures on advertising, and expenditures on R&D. If there are no sunk costs, potential entrants can use a hit-and-run strategy in which they enter an industry, undercut the price of the incumbents, reap the profits and exit before the incumbents have time to retaliate. In anticipation of entrants acting in this manner, the incumbents forestall entry by keeping price at average cost. The consequence is that, even in an industry that is highly concentrated, prices can be kept at or near competitive levels.

However, if sunk costs are present, firms entering an industry are unable to exit again without losing a portion of their investment. As a result, hit-and-run strategies are much less profitable and incumbents are able to keep price above average cost. Thus, with sunk costs, markets are not contestable and market power is once again an issue.

VERTICAL STRUCTURE

In addition to becoming larger through growth and horizontal integration, the major agrochemical companies have restructured themselves in other ways. One of the major changes is increased vertical integration—for example, the combining of seed, chemical and biotechnology activities within the same firm. Although some companies initially tried to include pharmaceutical and biotechnology components within the life science model, practically all of them have now divested their agricultural operations. A further element of this internal restructuring is the increased use of strategic alliances.

Many countries have had their own local seed companies that have over the years developed seed for a specific geographical market and operated sales and distribution systems. The major biotechnology companies are increasingly purchasing these seed companies as a source of seed material in which to insert the genes for herbicide or insect resistance. As an example, in 1997, Monsanto acquired a 30% share of the Brazilian corn seed market with the acquisition of Sementes Agroceres. With its 1998 purchase of Cargill's international seed division, Monsanto now controls over half the Argentine maize seed market. In 1998, Dow AgroSciences acquired Morgan Seeds, Argentina's second largest corn seed company, and Brazil's Dinamilho Carol Productos Agricolas, another key South American corn seed company. Phytogen (majority owned by Dow Agrosciences) acquired a major cottonseed breeding programme in the Chaco province of Argentina. In 1998, Mexico-based Empresas La Moderna (ELM) bought two South Korean vegetable seed companies and Nath Sluis (agricultural biotech company) of India (Action Group on Erosion, Technology and Concentration [ETC], 2000).

FACTORS THAT AFFECT HORIZONTAL INDUSTRY STRUCTURE

Numerous factors are behind the wave of horizontal mergers and acquisitions. Some of these factors are common to all industries and have no specific link to the seed, chemical and agricultural biotechnology industries. Thus, the mergers and acquisitions in the seed and chemical industries are at least in part a result of the need to consolidate costs and rationalize industry capacity, a desire by the management of the firms involved to extend their sphere of influence, and a wish by some firms to pre-empt other firms from taking over valuable assets. MacDonald (2001) provides an overview of concentration levels in a number of US agricultural sectors, many of which have seen higher concentration in recent years.

There are at least two explanations for increasing concentration, however, that pertain specifically to the seed, chemical and agricultural biotechnology industries: (a) sunk costs, and (b) escalation strategies.

SUNK COSTS

Two sources of sunk costs appear to be important in determining industry concentration: (a) intellectual property rights and R&D expenditures, and (b) the regulatory requirements that governments have introduced before the products of the R&D activity can be marketed.

On the empirical side, Ollinger and Fernandez-Cornejo (1998) examine sunk costs and regulation in the US pesticide industry. Using data over the 1972-89 period, they find that research costs and pesticide regulation costs negatively affect the number of companies in the industry, and that smaller firms are affected more strongly by these costs than are larger firms. Research and regulation costs also encourage foreign-based firms to expand into the US market and to force less profitable innovative firms to exit the market. Ollinger and Fernandez-Cornejo also point out that their results on the impact of regulatory costs generally match those found in other industries.

These empirical results can be linked to theoretical arguments concerning sunk costs and the economies of scale and scope that sunk costs tend to create. As Sutton (1991) outlines, the presence of sunk costs means that for firms to be profitable, price needs to be raised above marginal cost, typically by reducing the amount of competition (*i.e.*, the number of firms). Sutton identifies two types of sunk costs: exogenous and endogenous. Exogenous sunk costs are those beyond the control of the firms in the industry. Regulatory costs are a good example of exogenous sunk costs. In contrast, endogenous sunk costs are firm-level strategic variables. Advertising and R&D are examples of endogenous sunk costs. The rest of this section examines the connections between intellectual property rights (IPRs), R&D expenditures, regulatory costs, and economies of scale and scope.

Economies of scale and scope are major factors in propelling industries towards concentration. Economies of scale exist when average costs fall as more output is produced; economies of scope exist when the total cost of producing two outputs together is less than the cost of producing the two outputs separately.

Because economies of scale and scope mean that larger and diversified firms have lower average costs, there is clearly an incentive for firms to get large. Indeed, those that do not get large are vulnerable to being driven out of the market by larger and more cost efficient firms. Of course, there is a limit to how large firms can get. Although development and production costs may fall with an increase in the size of the firm, other costs—particularly those associated with administration—rise. Nevertheless, economies of scale and scope clearly create pressures for consolidation.

Economies of scale and scope are created as a result of investment in non-rival goods. Unlike rival goods—such as materials, labour, and energy—which can only be used in one place, by one person and at one time, non-rival goods can be used in more than one place and by more than one person, all at the same time. This feature of non-rival goods—that they can be used over and over again—means that output can be increased without having to increase all inputs. Consequently, economies of scale and scope are created. Much of the expenditure on non-rival goods—examples include R&D and regulation expenditures—can be considered a sunk cost.

Intellectual property, of course, is a good example of a non-rival good. Indeed, ideas generally are considered non-rival goods. It is widely believed that many of the high-technology industries, including the biotechnology and information industries, are subject to increasing returns. Romer (1990) postulates that this is a result of the distinction between physical goods and ideas. Ideas are not scarce, and as such, any industry based primarily on the trade of intellectual property will not face diminishing returns in their primary resource, the idea. The protection of intellectual property, through IPRs, is designed to encourage innovation and to provide incentives for the development and diffusion of new products and technologies.

An example illustrates the connection between intellectual property and economies of scale and scope. Suppose a biotechnology firm has some intellectual property, such as a particular gene that has been isolated. This intellectual property can be used over and over again as the firm expands its activities. For instance, if the company wishes to develop seeds for a new crop, it will not have to invest again in the research that isolated the gene. Although the development of a new seed will require additional lab and greenhouse space, labour, and materials, the expenditure on the technological advancements does not have to be made again.

A similar result occurs if the firm needs to invest substantial amounts of money in obtaining regulatory approval for a seed—although the production of additional units of the seed will require additional costs, the regulatory expenditures do not have to be made again. Once again, large companies typically have an advantage, because they are able to spread the costs of obtaining regulatory approval over more output. Thus, the greater the regulatory requirements in an industry, the more concentrated the industry is expected to be.

Intellectual property may also create economies of scope. Once a specific gene has been isolated—for instance, a gene that confers a resistance to a particular herbicide—this gene can then be put in a number of crops. Once again, the production costs of a number of products together will be less than if the products were produced separately.

To recap, R&D expenditures and regulatory costs are both sunk costs and a source of economies of scale and scope. Because economies of scale or scope mean that larger and more diversified firms have lower average costs, there is clearly an incentive for firms to get large. Indeed, those that do not get large are vulnerable to being driven out of the market by larger and more cost efficient firms. These theoretical results are supported by empirical observations in the US pesticide industry.

ESCALATION STRATEGIES

An escalation strategy is one in which a company spends large amounts on R&D and engages in mergers and acquisitions to achieve a dominant role in the market—that is, the firm tries to leapfrog its competitors to become the dominant firm. Escalation can be a profitable strategy when there is a high degree of substitutability with competitors' products on the demand side, and there are scope economies on the supply side. Both of these factors are present in the agricultural biotechnology industry.

The isolation of a gene that provides particular advantages and which can be inserted into a number of crops means there are scope economies on the supply side. There are also clear supply side scope economies associated with the enabling technologies that are required to use these genes. And on the demand side, herbicide and insect resistant seeds and the accompanying chemicals are clearly a substitute product for traditional seeds and herbicides and pesticides. Firms engage in escalation strategies in a number of ways. Although internal growth through R&D is one way, the more common method is through consolidation via horizontal mergers and acquisitions. As the theory predicts, the combination of demand substitutability and supply side scope economies present in the seed and chemical industry appears to be linked with escalation strategies.

Monsanto has clearly followed an escalation strategy, with Dow, Novartis, and others following similar strategies of making acquisitions and spending significant amounts on R&D.

BIOTECHNOLOGY FOR SOCIETAL DEVELOPMENT IN INDIA

The application of biotechnology for societal development has always received special attention since it benefits socially and economically disadvantaged sections of the society. Through this programme a number of projects are implemented for economic and technical empowerment of the beneficiaries. Efforts are made through training, demonstration and appropriate R&D inputs for providing tailor made solutions for location specific problems of SC/ST, women and rural population using modern biology and biotechnology. The following pages provide the highlights of the achievements.

PROGRAMMES FOR SC AND ST POPULATION

Biotechnology-based Programme for SC/ST population has benefited around 15,000 people directly and about 45,000 indirectly through 54 projects during the year. The projects have been supported in the areas of cultivation of medicinal and aromatic plants, biofuels, production of biofertilizers and biopesticides, organic farming, aquaculture, mushroom cultivation as well as human healthcare interventions etc. The universities, public funded institutions, Krishi Vigyan Kendras and voluntary/non-governmental organizations have been involved in the implementation of the projects. The target groups have received the benefit of hands-on training and field demonstrations.

Vermicomposting: MR Morarka GDC Rural Research Foundation Jaipur implemented a project in Jalgaon district of Maharashtra, to benefit 2600 SC/ST target population on economic utilization of waste, agriculture residues and cow dung. Over 1780 MT of waste, agricultural residues and cow dung was utilized for vermicomposting through establishment of 28 community facility centres.

Beneficiaries are utilizing vermicompost on crops like onion, sugarcane, cotton, banana, papaya and wheat. Some farmers switched over from conventional crops to cash crops such as colorful capsicum, gerbera as well as inter cropping of safed musli, turmeric and ginger in banana plantation. Farmers have been trained to use bio-fertilizer, foliar sprays and mulching to use the agro-waste for fertility and management. Marketing linkages have been established with "Organic Shoppe", through which farmers are getting 10-20% premium for their organic produce. About 20 entrepreneurs have taken up vermiculture as commercial activity. extension Material in the local language was distributed to the farmers.

Mushroom cultivation: A project on promotion & production of essential oil, mushroom and vermicompost was implemented by RRL, Jorhat for revenue generation of the ST population of Arunachal Pradesh. Motivation and demonstrationcum-training programmes have been conducted in 13 villages of five districts on cultivation of citronella, production of mushroom and vermicompost. Demonstration units have been established at RRL Branch Naharlagun farm. Mushroom spawn, earthworms and citronella saplings were distributed to the beneficiaries. Several beneficiaries established vermicompost units on commercial basis and are selling vermicompost. The beneficiaries at Ziro, Jairampur, Doimukh, Manmao, Hetlong and Tengman have established oyster mushroom cultivation units. So far, 10 acres of land have been planted with citronella and 7 acres of land with patchouli at Sonajuli village of Papum Pare district.

Medicinal plants: A project on strengthening health care practices and conservation of medicinal and nutritious plants was continued at Herbal Folklore Research Centre, Tirupati for implementation in 100 villages of two Mandals in Andhra Pradesh. Training workshops were organized with the help of Yanadi tribal people for community health volunteers on identification and sustainable harvesting methods of 20 species of locally used medicinal and aromatic species. The reputed healers were encouraged to prepare medicines needed for PHC from the identified plants. About 6000 seedlings were distributed to the beneficiaries for planting in kitchen gardens. Green health kits (samples of herbal medicines) and educational material in local language containing identification, cultivation techniques and preparation of medicines with their therapeutic properties were distributed to the trainees.

A project on creating awareness and promotion of medicinal plants for the production and marketing was implemented by RRL, Jorhat to benefit ST population of Arunachal Pradesh. Awareness and training programmes were organized on the importance, scientific cultivation and marketing of Piper longum, Kaempferia galanga, Andrographis paniculata, Stevia rebaudiana, Rauvolfia serpentina and Mentha piperita. Three nursery-cumdemonstration centers were established at Roing (1000 ft msl), New Bomdila (8160 ft msl) and Ziro (4350 ft msl) and 19000 saplings were distributed to ST farmers/NGOs for introduction and propagation in their fields. At high altitude, villagers were advised to cultivate Aconitum heterophyllym, Coptis teeta and Picrorrhiza kurroa. Beneficiaries have stopped collecting wild Homalonema aromatica, an important but rare aromatic plant of Arunachal Pradesh, and have started its cultivation in their farms. Ten community nurseries have been established with the help of Vivekananda Trust, Roing at ten villages of Lower Dibang Valley District. Twenty farmers were trained on cultivation and post-harvest management of medicinal plants and marketing tie-up has been established. Around 1266 local ST people participated in the awareness programmes.

A project on organic cultivation of medicinal and aromatic plants for income generation and sustainable development in Uttarkashi District of Uttaranchal was undertaken through Forests Conservation Programme of World Wide Fund for Nature, New Delhi. Demonstration-cum-Resource Material Production Centre established for training on vermicomposting and cultivation of species. Construction of poly house, vermi-composting units and planting of seeds of Rauvolfia serpentine completed. Awareness and motivation camps were organized for the farmers. Extension material brought out in the local language for cultivation of Rauvolfia serpentina, Coleus barbatus and Pelargonium graveolens were distributed to the farmers.

A project on developing skills and economically sound vermicompost units utilizing distillation waste of Mentha arvensis (menthol mint) was implemented for the benefit of SC/ST communities of Lucknow and Barabanki Districts of Uttar Pradesh. Around 200 farmers were trained on menthol mint nursery raising, transplanting, distillation and practices followed in vermicomposting. Demonstrations were carried out in farmers' field using high yielding variety 'Kushal'. Farmers trained on organic farming started vermicompost production using distillation and other agro-waste. The yield data obtained from the demonstrations showed the possibility of reduction of chemical fertilizers by 75% without affecting the yield.

BIOFUELS

A project was continued on cultivation of Jatropha in arid lands of Namakkal District in Tamil Nadu for production of biofuel and other products for SC/ST community by Centre for Research in Social Sciences, Coimbatore. A high tech nursery was established and plantlets of Jatropha were raised. Training programmes were organized for 223 members on cultivation of Jatropha in 31 acres of land. About 30,000 plants were supplied for cultivation. Awareness camps, training and demonstrations were conducted on Jatropha cultivation, oil extraction and soap making. SHG groups were trained in nursery techniques. Women self help groups were trained in various activities of jatropha cultivation and soap making. Extension material bought out in the local language was distributed to the beneficiaries.

Biocontrol agents: A project was implemented on development of eco-preneurship among the coastal community of Pondicherry by MS Swaminathan Research Foundation, Chennai. Sixteen SHGs involving women and youths were formed in Ariyankuppam, Nonankuppam, Kakayanthope and Veerampattinam villages. Demonstrations and trainings were conducted to educate trainees on basic aspects of IPM, importance of biocontrol agents, production and marketing of Trichogramma, importance of human nutrition, role of edible mushrooms and production & marketing of oyster mushroom. Micro-enterprise was developed for mushroom and Trichogramma production. Technologies were disseminated on biopesticide (Trichogramma), oyster mushroom production and vermicomposting to recycle the spent waste of oyster mushroom. Extension material brought out in the local language was distributed to the villagers.

AQUACULTURE

A project was undertaken on enhancing livelihood of SC/ST people in Orissa and Tamil Nadu by MS Swaminathan Research Foundation, Chennai for introducing various aquaculture enterprises to 250 participants in backyard ornamental fish culture and carp seed production. Fingerlings produced were distributed to the entrepreneurs for stocking. Various aspects on ornamental fish breeding, carp nursery and gold fish culture were disseminated to the self help groups, which helped them in earning a revenue of ₹ 200/month/household and an income of about ₹ 3200/from carp nursery and ₹ 8500/from vegetables and flowers as bund crops. Community ponds were effectively utilized by stocking with rohu, catla, mirgali, Chinese & Indian carps and marketable size fish was harvested. Technologies transferred to the entrepreneurs on community pond aquaculture, carp nursery and ornamental fish culture and extension material in local language were distributed to the beneficiaries.

A project on utilization of integrated marine living resources by SC youth from Tuticorin region of Tamil Nadu was undertaken by College of Fisheries, Tuticorin. Lobster fattening and crab culture techniques were standardized and demonstrated to the beneficiaries, who have grown lobsters in cages for about 4 months and produced 250-350gms size for marketing. The beneficiaries were helped in availing credit facilities and bank loan. In the project around 120 people of target population have been trained. A manual on crab and lobster fattening technology was brought out for the distribution to the beneficiaries.

A project on freshwater prawn culture was implemented in village Naubasta Kalan in Allahabad District by University of Allahabad. Demonstrations were conducted on prawn hatchery technology and standardized methods for biocontrol of prawn disease for integrated giant freshwater prawn (Macrobrachium rosenbergii) culture. Trainings were conducted on scampi culture, breeding and seed production and microbial inputs were utilized for feed silage preparation using indigenous technology. Apiculture activity was introduced and disseminated to the target population and extension material was distributed to the beneficiaries.

GENETIC COUNSELING

Tribal population of northeast region was screened for anaemia and nutritional deficiencies through a programme implemented at Vivekananda Institute of Medical Sciences, Kolkata. The tribes of Arunachal Pradesh, Assam and Tripura displayed high incidence of anaemia. Awareness programme was organized for school children on inherited anaemia in pre-marital and pre-natal population. More than 1500 families from 80 villages were educated on anaemia, thalassaemia and hemoglobinopathies and counseling was provided to them. Leaflets were distributed for awareness of anaemia with special reference to Thalassaemia in local languages.

A project on molecular epideomiology of alcoholism was implemented jointly by Centre for Cellular and Molecular Biology, Hyderabad and Tribal

Research Centre, Udagamandalam, Nilgiri District to study Kota tribe population in Nilgiri hills. General health awareness camps were arranged in seven villages for periodical medical check up. Blood samples were collected from 620 individuals and clinical data on health profile revealed that the disease morbidity is more for hypertension, heart related problems and diabetes. Genealogies indicated that the alcohol drinking habit is running in the families for the past two to three generations. Study undertaken on the health awareness camps and medical camps were found beneficial in counselling the tribal population.

FLORICULTURE

Over 180 women from Shillong, Itanagar and Chandigarh were trained in cultivation of commercially important and hybrid species of orchids. A total of 118 women of Shillong have raised 6 species *viz.* Dendrobium fimbriatum var. oculatum, D. longicornu, D.lituiflorum, C. pendulum, C. aloifolium and C. giganteum in their backyard. These plants have started blooming. The beneficiaries have started selling their harvest in local market. Through S&T Council, Bhopal 200 women were trained in cultivation of gladiolus, marigold, mogra, rose and tuberose.

FOOD PROCESSING

At College of Horticulture, Vellanikara, Thrissur, 88 women were trained in various aspects of entrepreneurship in preparation of cocoa products including molding, packaging and marketing. Twenty-two women have already set up their production unit and are selling their products in local market. In another project at Fisheries College, Tuticorin, 500 women were trained in cook-chill technology for seafood and various parameters for quality assurance *viz.* estimation of moisture content, bacteria load, presence of bacterial species, the nutritive value, shelf-life etc. The trained women were grouped in 48 SHGs for establishing market for seafood products.

At the Central Food Technological Research Institute, Mysore data on biochemical and microbial analyses for nutritionally important factors like protein, amino acids, isoflavones, genistein, diadzein and methods for improvement of shelf life were collected on for 'Tofu' preparation. Also a byproduct of Tofu preparation, Okara, was found to be rich in protein and amino acids has been used as a food supplement. In another project at CFTRI, 17 women were trained in preparation of various food products of high nutritive value, methods of removal of anti-nutritional factors like phytic acid and trypsin inhibitor found in soya bean, better texture and improved shelf life. The food items included fermented products of soya bean and fruit pulp. Also 34 women were trained in spirulina production and its application as health food. Through CFTRI, technologies for rural bakery are being disseminated to rural women in three villages.

POULTRY AND LIVESTOCK FARMING

The 'Action Research on Health and Socio Economic Development', Bolangir, Orissa, identified 100 women and trained them in poultry rearing. The beneficiaries were provided with parent birds of Carigold (dual purpose) and Vanaraja varieties. The beneficiaries were able to earn ₹ 1000/per month through the sale of eggs and birds. The beneficiaries were also trained in feed formulation using local ingredients, management practices and entrepreneurship skill. The trained women formed SHGs and are approaching a bank for setting up their poultry farms. The market is established through state owned apex cooperative known as OPOLFED.

Through three projects supported at Tamil Nadu Veterinary & Agricultural Sciences University and its Extension Centres at Namakkal, Chennai and Tiruchirapally, over 1200 women were trained in poultry farming including feed formulation, farm management / health care and preparation of various food items such as quail egg pickle and ready-to-eat food products etc.

Aquaculture: Under a project at the North Eastern Hill University, Shillong, techniques were standardized for induced breeding for common carp and 4,00,000 fish seeds were reared up-to fries for distribution to the beneficiaries. Selected 92 women were trained in fish farming techniques including water quality testing, pond maintenance, induced breeding, harvesting and feed formulation using slaughter house waste. They were provided with fish fries for rearing in their own pond.

Over 470 women from Bhopal, Shimla, Solan and Bhubaneshwar were trained in polyculture fish farming. Interested women were also provided advance training in composite carp culture, integrated fish farming with duck/ poultry, pond management and value added products including fish processing. These women are selling their products in local market and are earning ₹ 3000/per harvest/family.

Considering the opportunities in ornamental fish, a project was supported at the University of North Bengal. The identified women, 57 from Darjeeling and 54 from Jalpaigudi district were trained in rearing and breeding of guppy (Poecilia reticulate) in cemented cisterns and angel (Pterophyllum scalaer) in ponds. The women are being trained in management skill and marketing.

ORGANIC FARMING / VERMICOMPOST

At the Sri Sathya Sai Institute of Higher Learning, Anantapur Campus, Andhra Pradesh, neem varieties with higher level of azadirachtin were identified in 5 villages. Thirty trees were selected and 40 women from two villages were trained in fruit collection, drying and storage. Specialised training on maintenance of neem processing gadgets, processing of neem fruits for production of cake, oil or powder, maintenance of aza-rich plantlets, hardening, harvesting and application of neem based biopesticide was provided to five resource persons At the University of Agricultural Sciences, GKVK Campus, Bangalore, five villages of Bangalore and Kolar were surveyed for the availability of organic waste. The waste materials were mostly sugarcane trash, coconut coir pith and vegetable wastes. Demonstration programmes for 20 to 25 farmwomen from each identified village were organized. Six interested women were given onfarm training. The technique of Trichoderma enriched vermicomposting was perfected and demonstrated at University campus.

At MSSRF, Chennai, phosphate-solubilising local strains were collected and screened. Simultaneously several awareness programmes on application of biofertilizer were organized for the villagers of Rediarchatram block. About 300 interested farmers were trained in large-scale production of biofertilizer.

Under the project support at Dangoria Charitable Trust Hyderabad, apart from training women in processing various items of food, 100 women from 13 villages were trained in production of vermicompost, neem-based biopesticide and their application in organic farming, About 100 vermicompost units were set up using the ring method. The beneficiaries produced over 130 quintals of compost. The beneficiaries were encouraged to raise nurseries of drumstick, papaya and creeper spinach at their backyard using vermicompost. The beneficiaries were able to sell excess vermicompost and planting material.

HORTICULTURE

Through the Tamil Nadu Agricultural University, Coimbatore, 54 progressive farm women were identified for training on cultivation of hybrid rice. The women were provided with quality seed their cultivation over an area of approximately four acre. The yield for the first harvest (December to May) was recorded at 2000 kg/ha. The beneficiaries were able to earn an additional income of ₹.20,000/per acre/year through three crops in an year. At Supi farms, Uttaranchal non-traditional vegetable crops were grown on farmers' land in a small way. An herbal garden was established. Incidence of pests to various crops was evaluated and farmers were trained on production of vermicompost and application of neem based biopesticide in horticultural crops.

MUSHROOM CULTIVATION

In an attempt to empower rural women through transfer of mushroom cultivation technology 93, women were trained at Himachal Pradesh Krishi Pradesh Vishwavidyalaya, Palampur and 750 at Tropical Botanic Garden & Research Institute, Kerala in mushroom cultivation. In addition, 15 women were trained in spawn production at TBGRI. The beneficiaries were given 15 days training in cultivation of Oyster and Agaricus species using wheat straw. The first harvest of about 1000 kg was sold @ ₹ 16/kg. The beneficiaries were also trained in various recipies of mushroom. Market linkage has been established for their products. Under the project support at Dangoria Charitable Trust Hyderabad, apart from training women in processing various items of food, 100 women from 13 villages were trained in production of vermicompost, neem-based biopesticide and their application in organic farming, About 100 vermicompost units were set up using the ring method. The beneficiaries produced over 130 quintals of compost. The beneficiaries were encouraged to raise nurseries of drumstick, papaya and creeper spinach at their backyard using vermicompost. The beneficiaries were able to sell excess vermicompost and planting material.

SERICULTURE

A total of 550 women were trained for various activities including the organic cultivation of mulberry & rearing of cocoon and value added products from waste at the Central Sericultural Research & Training Institute, Mysore and at Vidyasagar University, Midnapore, West Bengal. The beneficiaries are getting an additional income through the sale of value added products and surplus vermicompost.

HEALTHCARE

Kidwai Memorial Institute of Oncology, Bangalore, Karnataka organized over 50 awareness camps on Carcinoma cervix to educate 2071 women for its prevention. Blood samples collected from 1830 women based on their medical history, 1812 were found negative and 18 positive cases were referred for treatment. A project on mental retardation including awareness generation among rual women, screening for early detection and genetic counseling was supported to Manovikas Bio-Medical Research. Over 7,000 families from three Gram Panchayats were surveyed. Ninety-one of these showing signs of mental retardation were shortlisted. The families were informed about the possible risk factors in future. Analysis of blood samples from these for genetic carrier showed only 4 cases of low level of fragile X chromosomes.

A project is supported at Saraswati Dental College, Lucknow for dental fluorosis. Water samples from 12 villages around Lucknow and Kanpur were analysed for fluoride content. Health camps were organized in 4 villages with a team of dental surgeon. A total of 2860 people were examined for dental fluorosis. The severely affected persons were advised for regular dental examination at SGPGI, Lucknow. They were also provided with calcium and cholicalciferol supplement. The blood samples from the severely affected boys and girls within the age group of 15 years and pregnant women were examined for their blood calcium level. Under another project supported at Mahavir Hospital And Research Centre, Hyderabad, Andhra Pradesh, blood lead levels in children and adults including men and women were measured to study environmental genotoxicity in women and children occupationally exposed to lead, a heavy metal.

IPRS AND INVESTMENTS IN PLANT BIOTECHNOLOGY

The implicit rationale for IPR protection is that they promote investments in plant breeding and bio-engineering. This hypothesis is contested by section of analysts. Nevertheless it is a fact that the structure of IPRs do influence investments. IPRs induce their own pattern of innovations. The time taken for R&D to fructify as innovations is a crucial aspect guiding investments.

Often IPR regimes play a key role in influencing the time pattern of innovations. It is a well-known fact that conventional plant breeding methods are dilatory and time consuming. It requires 7 to 8 generations of repeated breeding to reduce heterozygosity of new genotypes. As Vossen (1985) observes Arabica Coffee subject to three to four cycles of breeding still display heterozygosity and it is not until 20 to 25 years (spanning 4 to 5 generations) that one arrives at true breeding varieties. This time lag is even more pronounced for the Arabusta coffee variety (inter-specific hybrid of Arabica and Robusta Coffee) where segregation of unfavourable characteristics could still occur even after several generations of backcrossing and selfing.

Another interesting example is the one cited by Evenson (1991) about the long gestation period in regard to rice varieties in India. According to Evenson, India released 306 rice varieties for planting during the period 1965-1986 after performing 20,000 crosses over a period of over 15 years since 1950. It must be understood that this long process of backcrossing was preceded by considerable time spent on location and identification of individuals plants with requisite traits. The speed or pace at which plant varieties qualify for being released for planting have economic implications on investment behaviour in seed/plant breeding industry. Asexual modes of plant breeding which include plant genetic 'engineering' and transgenic techniques could obviate the long gestation period associated with pre-release preparations encountered conventional breeding.

It is a different matter than biosafety concerns may delay releases of these varieties. The time lag in release of plant varieties fulfilling the criteria of novelty, distinctness, uniformity and stability (or alternatively the requirements of novelty, inventiveness and commercial application) raise fundamental issues from the view-point of capital investments. The time lag in varietal release is potentially least in the case of transgenic and non-sexually produced varieties while for traditional or classical methods the time lag can be greater.

The UPOV 1991 based plant IPR regimes could induce longer time lags as compared to the situation when an IPR regime is based on the 1978 version of the UPOV. This is due to the fact that the minimum genetic distance concept implicit in UPOV 1991 precludes 'close cousin' varieties from acquiring protection rights, unless the techniques of genetic distance determination by themselves are inadequate. UPOV 1978 would facilitate capital investments to some extent as the release of plant varieties is periodically more frequent.

Even under the plant patent regimes, investments could bring about short-term returns, though the volume of investments will have to be larger on account of capital intensity of advanced plant genetic engineering technologies. By comparison, UPOV 1991 regime is least conducive to capital investments as plant variety releases can be painfully slow under this regime. Therefore sustenance of conventional plant breeding under UPOV 1991 regimes, can only be ensured by infusion of low discount capital.

Paradoxically it is the reduced availability of low discount public funds for plant breeding that has induced countries like India to go in for Plant Varieties Protection Legislations modelled on UPOV lines. The implicit assumption in undertaking this step is that private funds for R&D in plant breeding will 'crowd in', if protection plant varieties rights are accorded. But the reality is that in liberalised deregulated developing economies, private investments have 'crowded in', in short return promising avenues.

Consequently classical plant breeding activities will suffer for want of funds in the UPOV 1991. In other words, UPOV 1991 modelled national legislations will not necessarily hold attraction for private sector investors in R&D projects involving classical plant breeding as compared to say, UPOV 1978 or the Plant Patent regimes. Rather private investments would be catalysed better by *sui-generis* plant protection legislations, which partake of plant patent elements and focus in the main on hybrids and genetically engineered varieties.

Given the linkages between plant breeding and the seed industry these perverse investment behaviour trends can produce major economic consequences. The Seed Laws in various countries, both advanced and developing, are getting to be progressively oriented towards restricting market circulation to varieties that are conferred with plant breeder or patent rights. The UPOV guidelines on DUS are used world wide not only as the basis for establishing varietal distinctiveness and descriptions but also for seed certification purposes. The DUS guidelines are employed for recognising and registering not only the 'basic' seeds used for multiplication of 'pedigree' seeds but also for certification of the pedigree seeds themselves.

BIOTECHNOLOGY SYSTEM IN INDIA

The government plays an important role in building the National Innovation System (NIS) by setting up institutions (public research institutions such as universities, labs etc), developing policies that support innovation and providing regulatory frameworks that can have a considerable influence on sectoral development.

In the context of the developing countries, the role of the government becomes more prominent as the majority of the resources for the scientific research and development are being channelised or governed by the government. In case of India, nearly 80% of the gross expenditure for research and development is being incurred by the Government while in case of the industrialized nations such as Japan, USA and UK the Industry spends nearly 75%, 64% and 44% respectively. Though there has been a significant increase in the country's Gross National R&D expenditure in absolute terms from ₹.3974 Crore in 1990-91 to ₹.16199 Crore in 2000-01 to ₹.18000 Crore in 2002-03, the share of the government in national R&D expenditure has remained unchanged. However, it is interesting to note that internationally all the governments perform the in-house R&D function, including even the highly private-sector-oriented governments such as the US and other OECD countries. A study of the Biotechnology innovation system of the four global leading countries namely USA, UK and Japan and Germany further supports the active involvement of Government R&D in furthering the growth of the Biotechnology sectoral innovation. Biotechnology is being characterised by multidisciplinary fields, high dependence on science and are emerging from industries having large R&D base, high degree of qualified manpower and having close links or interaction with public research institutes as well as universities and thus faces a broad set of different policy interventions by the government underlying the National Innovation System (NIS).

The National System of Innovation (NSI) perspective is a convenient framework to understand the process of innovation occurring in an economy and especially within the manufacturing sector. It focuses on the country-specific factors influencing the process of innovation and economic growth. The NSI approach has evolved over the years both analytically and empirically with the contributions from Freeman, Lundvall and Nelson, Edquist (1997, 2002). NSI has became popular as an analytical framework to study innovation at the national, regional and even Sectoral level.

Though there are several definitions of NSI but that best fit in to the present context is that of Metcalfe (1995). NSI is defined by Metcalfe (1995) as a "set of distinct institutions which jointly and individually contribute to the development and diffusion of new technologies and which provides the framework within which governments form and implement policies to influence the innovation process. As such it is a system of interconnected institutions to create, store and transfer the knowledge, skills and artefacts which define new technologies".

NSI framework is comprised of various components such as the government, independent research institutes, firms and higher education system etc and their interrelationships. Innovation or government policy instruments and institutions play a very important role in cementing the relationship among the various components for the success of NSI. An application of this framework shall help in analysing the systemic failure that hamper the generation of innovations. Public policy can be applied to correct such failure. There are several NSI studies on Biotechnology in the context of the developed world but the studies in the developing countries are rather limited. Biotechnology NSI in India has evolved over time and has been instrumental in creating infrastructure and capabilities in promoting biotechnology innovation in the country. There are 241 institutions and 321 firms engaged in various dimension of biotechnology in the country. Indian Biotechnology industry is now among the top 12 countries in the world and like its global counterpart, is dominated by the healthcare sector. The recently released "Draft National Biotechnology Strategy 2005" by the Indian Government is a step forward in the direction of promotion and development of biotechnology to withstand global challenges and yet, solve the local healthcare needs of the country. Studies on biotechnology in India by and a recent empirical analysis of the health innovation system of seven developing countries further supports the role of the Governments in providing the sustainable environment for the successful development of health biotechnologies.

According to Acharya *et. al.* government in developing countries is one of the key actors in prioritizing, implementing new technologies and promoting innovation to address their specific health and development needs. The governments of the countries where biotechnology innovation system has been successful in perform or support the following functions/activities such as knowledge support, commercialization of research, collaborative research, financial support, regulation, IPR and resolving ethical issues etc.

Thus it would be interesting to analyse the role of the government in the promotion of the biotechnology in the country with special focus on the healthcare sector.

Hence, the present study traces the role of government or national approaches in terms of policies, institutions and instruments in the promotion of biotechnology innovation system in India with a special focus on the healthcare biotechnology sector. A systemic framework similar to the EPHOITE (2003) project is being adapted to examine the direction of public policies, institutions and instruments in addressing the issues of knowledge support, commercialization of research, collaborative research, financial support, regulation, IPR and ethical issues in the area of healthcare biotechnology in Indian context.

The significant features of national biotechnology innovation system with focus on healthcare biotechnology with the help of various S&T indicators. The various government initiatives in terms of various policies, institutions and instruments in the framework of NSI while the lessons learnt along with the policy implications and challenges for the growth of healthcare biotechnology innovation.

BIOTECHNOLOGY AND INDIA AGRICULTURE

We need to recongnise that, just as the food reequirements of today's population of nearly 6 billion people could not have been met by the technologies of the 1940's, we cannot assume that current practices will feed the population of 8 billion expected by 2020. New approaches are needed in addition to the continued improvement of existing methods of crop and animal husbandry and food processing.

The strategic integration of biotechnology tools into India agricultural systems can revolutionise Indian farming and usher in a new era. Genetic engineering is clearly the most revolutionary tool to impact agricultural research since the discovery of genetics by Mendel. Prior to genetic engineering, the exchange of DNA material was possible only between individual organisms of the same species. With the advent of genetic engineering in 1972, scientists have been able to identify specific genes associated with desirable traits in one organism. For example, a gene from bacteria, virus or animals may be transferred into plants to produce genetically modified plants having changed characteristics. This method therefore allows mixing of the genetic material from species that cannot otherwise breed naturally.

Genetic engineering can be vital for an agrarian country like India. It can help in minimising the crop damage through disease and pest resistant varieties, reducing the use of chemicals, enhancing stress tolerance in crop plants, thus permitting productive farming on unproductive lands, etc. One could even extend the growing season of crops and minimise losses due to environmental factors on the one hand and increase the shelf life of fruits and vegetables, on the other, thus minimising losses due to food spoilage.

We can expand the market vista and improve food quality. Biotechnology can also produce plants that possess healthy fats and oils, possess increased nutritive value, and create a whole range of higher value feeds. Biotechnology has even the power of producing biodegradable plastics, edible vaccines, etc.

It is only through the blending of the 'gene revolution' with our experience in the 'geen revolution' that we can reach our goal of 'evergreen revolution' and also 'nutritional revolution' The advantage of the gene revolution is that it is relatively scale neutral, benefiting big and small farmers alike, It is also environment friendly.

Thus, it can be of great help to the smallest farmer with limited resources in increasing farm productivity through the availability of improved but powerful seed. It can also reduce a farmer's dependency on chemical inputs such as pesticides and fertilizers.

Modern biotechnology offers unlimited opportunity for enhancing genetic potential of crops and other commodities, management of biotic and abiotic stresses, bio-remediation and organic recycling. India is one of the main centres of agricultural biodiversity and its gene richness can greatly complement the developments in modern biotechnology. Likewise, new developments in GIS, remote sensing, and crop modelling, provide new opportunities for integrated management of natural resources. The revolution in informatics provides opportunities for sharing latest information for research and planning in a highly organised and efficient manner. With the globalisation of economy, the opportunities for value addition and post harvest management are also immense.

With clever blending of technologies, the India farmer can usher in new era of confidence and performance. Sir Francis Bacon had once said: *'It would be an unsound fancy to expect that things which have never yet been done, can be done except by methods which have never been tried'* In the new context, we have to review the new role of biotechnology. Also, it is in this context that we have to view the rapid advances that modern biotechnology is making, including the area of transgenics. Let us focus on some of the developments in this area, since considerable interest and controversy has been created around the world today in this exciting area.

MOVEMENT IN TRANSGENICS

The advent of genetic modification over the last two decades has enabled plant breeders to develop new varieties of crops at a faster rate than was possible using traditional methods, with huge potential for further beneficial development.

Tobacco was the first plant to be genetically transformed\, in 1983, with cereals beginning in 1990, but it is only recently that products such as GM soya have reached the market place in Europe, and, for a variety of reasons, given rise to concern & controversy.

The concern must be addressed seriously if society is to exploit the new technologies appropriately. We will address this issue rather extensively a little later. Whereas in India, we have been discussing and debating the issues, the rest of the world is galloping ahead.

Table. Ag Biotech Product Sales

$ Millions		**Sales**		**Annual Growth Rate**	
	1994	1997	2002	1994-97	1997-02
Transgenic seeds and plants	20	405	2,030	173%	38%
Animal growth horomes	80	225	405	31	12
Biopesticides	35	65	110	23	11
Other	107	180	340	19	14
Total U.S. Sales	$242	$875	$2,885	53%	27%

Development, field testing, and commercialisation of transgenic crops, is now recongnized by the international scientific community to be an essential strategy for food security. Fortyfive countries, including India, have conducted transgenic crop field trials up to 1997. The activity in the developing countries has been shown in Table below.

Table. Releases of Transgenic Plants in Developing Countries (by Species and Introduced Trait)

Species	Introduced Trait					
	Total Fied releases	Herbicide Resistance	Insect Resistance	Virus Resistance	Product quality	Others
Maize	46	29	16	1	3	5
Soya bean	27	25	—	1	1	—
Cotton	24	16	15	—	—	—
Tomato	19	—	2	1	16	—
Potato	13	—	1	6	—	6
Subtotal	129	70	34	9	20	11
Other species	30					
Total	159					

The emphasis has been mainly on traits for herbicide, insect, virus & fungal resistance, and on product quality with a limited number of genes. Out of a total 159 releases. Argentina leads the way with 43, followed by chile, Mexico, Puerto Rico and the Republic of South Africa, with 17-20 each. India is far behind with just a few releases. There are fears relating to transgenic plants in society. These need to be addressed, not by brushing them aside, but by careful scientific analysis and educational programmes. All biosafety protocols must be followed rigidly. Let us deal with this issue more exhaustively now.

PUBLIC CONCERNS

The potential of biotechnology as a method to enhance agriculture productivity in the future has been accepted globally. However, because of its revolutionary nature, risk and uncertainty may be created by the process of genetic engineering and by the resulting genetically modified products. This may result from the introduction of a new unrelated DNA sequence into a recipient organism. The introduced DNA may have some unexpected effects on the cellular processes of the recipient organism. In addition to this, introduction of antibiotic resistance genes, as selection markers, pose serious implications in public health especially in genetically modified plants that are directly used for food purposes.

Risks are also associated with the genetically modified plants that are released into the environment. The nature of interactions with other organisms of the natural ecosystems cannot be anticipated without proper scientific testing. For example, modified plants with enhanced resistance to pests or disease threaten to transfer resistance to the wild relatives, which will have implications for biodiversity and ecosystem integrity. These, and many more, doubts plague the minds of common people. Therefore, it is very clear that if the biotechnology potential has to be completely realized, it has to be done in an extremely responsible manner. At each step, proper testing and scientific data have to provided. Not only this, proper regulatory and policy mechanisms have to be put into place.

It is therefore, absolutely essential, that conscious efforts be made especially by the developing countries to initiate well-defined programmes for the development and regulations of genetically modified plants. To make this possible and to benefically, ethically and sustainably reap the benefits of biotechnology, there is a pressing need for scientists, researchers, policy makers, NGOS, progressive farmers, industrialists, and representatives of the government, to come together on a common forum in order to discuss the common concerns and find solutions to them.

Will modified plants transfer their introduced genes into wild relatives growing nearby? Will modified plants that produce new compounds disrupt the 'balance of nature' in some way? Could the planting of a restricted number of cultivators leave crop plants more susceptible to disease? Could the planting of a restricted number of cultivators lead to a reduction in biodiversity (of crop plants, weed species, insects and microflora in the fields in question)?

Will genetically modified plants be able to avoid the factors that regulate natural populations and thereby change the usual ' balance' between populations? Similarly, there are concerns about the social and economic effects. For example, some of these concerns are:

- How will the structure of farming (particularly in developing countries) be affected by biotechnology?
- How will patent laws affect traditional breeders 'rights (*e.g.* the right to save seed from one year to the next)?
- Will plant breeding be left increasingly in hands of a few companies, and if so, what effects might this have?
- Will some countries be 'plundered' for their genetic resources?
- We also have a set of ethical and moral issues that need addressing. In particular,
- The consumer has the right. What will consumers be told about the new food products?
- Is it acceptable to 'interfere' with nature through genetic engineering?
- Do we have the right not to use all means available to improve crop plants, especially when so many people are under or malnourished?

Then there are several regulatory issues. Among these, the three key issues are:

- Do current regulations give sufficient protection to farmers, consumers, those who have invested in, and those engaged in research?

- Is there sufficient international legislation to ensure environmental protection?
- Do current regulations compromise the competitiveness of biotechnology companies by being excessively restrictive?

The Royal society of London had appointed a group of experts to examine various aspects including the scientific evidence concerning the risk of transfer of genes from Genetically Modified (GM) crop plants to wild species and non-GM crops, the uptake of genes from GM food by the digestive system, and the current state of the regulatory system.

The experts concluded that the chances of gene transfer happening are slight, provided the regulatory processes are followed, but that this must be kept under consideration.

The experts also concluded that the uptake of genes via the food chain is not a new issue because genes (*i.e.* DNA) are normal constituents of the human diet. Many products from GM Plants, such as sugar prepared from GM tomato paste, are so similar that they are regarded as 'substantially equivalent' others, for example flour from GM soya, may contain a new gene or its product, although many of the purification processes involved in food production will destroy and DNA present in the raw material.

Some GM foods have been produced using an antibiotic resistance 'market' gene, which is a laboratory device designed to identify genetically transformed plants.

The Royal Society has shown concern about the use of such genes in food products and it suggested that any further increase in the use of such markers in the human or animal food chain would be undersirable. Gm crop plants have been produced to improve insect tolerance and virus resistance, and to include herbicide tolerance, so that transferred to non-target species of plants, and that the development of resistance by target pests is minimised, the regulatory authorities must be assured that any negative effects would be no greater than those resulting from conventional procedures; and that any long term effects on the environment and ecology would be closely monitored, with statutory restrictions in place to control marketing; and also that best practice advice was adopted by growers.

The reliance on a case by case approach to the legislation, may result in lack of analysis of the overall impact of the technology on agriculture and the environment, and of the long term effects of gm. Although mechanisms are already in place to regulate many individual aspects of GM technology, there is no means for looking at GM technology as a whole. As independent, overarching, regulatory body is needed to span departmental responsibilities, monitor the enforcement of existing or future regulations, and strengthen the guidelines to growers of GM crops, such as those specifying the isolation distances between Gm and non-GM crops. The body should also review and monitor the membership of advisory committees and regulatory bodies.

INSTITUTIONAL SETUP ANDBUDGETARY ALLOCATIONS IN INDIA

India is one of the first few countries, among the developing countries, to have recognized the importance of biotechnology as a tool to advance growth of agricultural and health sectors as early as in 1980s. India's Sixth Five Year Plan (1980-85) was the first policy document to cover biotechnology development in the country.

The plan document proposed to strengthen and develop capabilities in areas such as immunology, genetics, communicable diseases, etc. In this context, referring to the Council of Scientific and Industrial Research (CSIR), the document suggested to ensure coordination on inter-institutional, inter-agency and on multi-disciplinary basis, full utilization of existing facilities and infrastructures in major areas including biotechnology. Programmes in the area of biotechnology included, tissue culture application for medicinal and economic plants, fermentation technology and enzyme engineering for chemicals, antibiotics and other medical product development; agricultural and forest residues and slaughterhouse wastes utilization and emerging areas like genetic engineering and molecular biology.

The existing national laboratories under the S& T agencies, such as Indian Council of Medical esearch (ICMR) and Council for Scientific and Industrial Research (CSIR) had initiated several esearch programmes to fulfill the above plan objectives. At the top, an apex official agency *viz.* ational Biotechnology Board (NBTB) was set up in 1982, to spearhead development of biotechnology. The NBTB was chaired by Member (Science) of the Indian Planning ommission and had representation of almost all the S&T agencies in the country *viz.* epartment of Science and Technology (DST), Council for Scientific and Industrial Research CSIR), Indian Council of Agricultural Research (ICAR), Indian Council for Medical Research ICMR), Department of Atomic Energy (DAE) and the University Grants Commission (UGC).

NBTB was formed with the specific purpose of the identification of priority areas and for volving a long-term plan for the country in biotechnology as well as to initiate and promote uch activities as conducive for further development of various areas in biotechnology. The BTB issued the " Long Term Plan in Biotechnology for India" in April 1983. This document pelt out priorities for biotechnology in India in view of the national objectives such as self ufficiency in food, clothing and housing, adequate health and hygiene, provision of adequate nergy and transportation, protection of environment, gainful employment, industrial growth and alance in international trade. Later in 1986, NBTB graduated to a full-fledged government epartment called Department of Biotechnology.

At present in India, there are six major agencies responsible for financing and supporting esearch in the realm of biotechnology apart from other sciences. They are Department of cience and Technology (DST), Department of Biotechnology (DBT), Council of Scientific and ndustrial Research (CSIR), Indian Council of Medical Research (ICMR), Indian Council of griculture Research (ICAR) and University Grants Commission (UGC), Department of cientific and Industrial Research (DSIR). DST, DBT and DSIR are part of Ministry of Science nd Technology while ICMR is with Ministry of Health, ICAR with Ministry of Agriculture and UGC with Ministry of Human Resource and Development. DSIR is the funding agency for CSIR and both of them independently fund biotechnology related research programmes.

Allocations for all of these agencies have gone up in the last decade. Out of this, DBT is the only agency completely devoted to R&D in biotechnology. It is very difficult to estimate the total allocations for this sector per se from other aforementioned agencies as in some cases the allocations are not separately marked as allocations for biotechnology. One faces this kind of constraint especially with those organisations, which are focusing on technological solutions and are not committed for X or Y nature of technology. Thus separately accounting for biotechnology becomes difficult. It would probably become possible only if a detailed survey at the institutional level is undertaken. A broad idea about total allocations by major ndian funding agencies to Science and technology related projects and not necessarily to biotechnology alone. In case of UGC it gives a broad idea not only about S&T related projects but also about other research streams.

DEPARTMENT OF BIOTECHNOLOGY (DBT)

There has been a significant increase in Government of India's outlays for biotechnology over the past decade. Since the time of establishment, in 1986, the allocation for the Department has increased manifold. The budgetary allocations have gone up from ₹ 404 million in 1987-88 to ₹ 1,138 million in 1997-98 and by 2001-2002 it became ₹ 1,863 million. Though the current price allocation figures may not give a complete picture but the budgeted figure for 2002-03 shows a doubling of allocation to ₹ 2,356 million. In India the developmental allocations are generally made for five years under the National Five Year Plans. The Government has recently finalized the Tenth Five Year Plan (2002-2007). The Working Group on Science and technology for the Tenth Five Year Plan, constituted by the Planning Commission, has proposed an outlay of ₹ 20,750 million for the period 2002 to 2007. This marks a sharp increase of 234 per cent from the budgetary provisions made during the Ninth Plan period (1997-2002) which totaled at ₹ 6,215 million only. The Vision Statement for the Tenth Five Year Plan enumerates the proposal for human genome sequences, proteomics, structural biology and bioinformatics.

The DBT has taken special precaution to find crucial balance between different sections of society as far as technology absorption is concerned apart from promoting industrial development. It is supporting low-cost biotechnology adoption programmes for socially backward communities. The programmes include vermi-composting, use of organic manure, silk-worm rearing, mushroom cultivation, etc. Some training cum demonstration programme are also being supported for them. Efforts are also on for gender mainstreaming. The DBT has launched 11 projects for women in the areas of waste management, bio pesticides, bio fertilizers, floriculture, fish farming for poor women in the rural areas. At the Golden Jubilee Biotechnology Park for Women, industrial modules have been allotted to women entrepreneurs for setting up units in the above-mentioned areas. This Park was inaugurated in 1998 near Chennai the capital of a southern Indian state Tamil Nadu. The Park is located in Siruseri, adjacent to Information Technology Park in an area of about 20 acres. It would have some central facilities for the entrepreneurs for technology resourcing, training and marketing. This is a first unique project of a joint effort by the central and state government in the realm of biotechnology.

CHANGING ROLE OF DBT

In recent times, DBT has taken up a proactive role in promotion of industry. DBT proposed a single window application-processing cell as part of a new regulatory system for the domestic biotechnology sector. The move formed part of the recent recommendations on biotechnology sector made by the Confederation of Indian Industry (CII). Besides recommending setting up of a single window application-processing cell at DBT, CII had also suggested a fixed time frame of 150 days for clearing new biotech proposals.

In this regard, CII had recommended a process whereby a new application would be sent by the single window agency to the Review Committee on Genetic Manipulation (RCGM), which in turn would be required to submit a scientific evaluation report within 60 days of receiving the applications. This report will be submitted to the relevant approval committee, identified by the end product category. For example, in case of agricultural products, it would go to the Genetic Engineering Approval Committee (GEAC), in case of pharmaceutical products to the Drugs & Pharma Approval Committee (DPAC), and in case of food products to the Biotech Foods Approval Committee (BFAC). The GEAC/ DPAC/ BFAC would be required to accord approval or rejection within 90 days of receiving the evaluation report from RCGM. In case of rejection of the application, the applicants will also have the right to appeal to the concerned approval committee. CII had also suggested that any additional information required by RCGM for completeness of the application form would have to be called for within 30 days of receipt of the application. Besides DBT, the recommendations were submitted by CII to 14 other agencies, including seven ministries. Most of the recommendations submitted by CII have been accepted by DBT.

The Council of Scientific and Industrial Research was established in 1942. It is India's largest research and development organization. It has 40 laboratories and 80 field stations/extension centres spread over the length and breadth of the country. The total allocation for CSIR in the year 2000-01 was ₹ 9120 million, which is 13 per cent higher from 1999-2000 (₹ 7940 million). CSIR's Centre for Cellular and Molecular Biology (CCMB) incubated India's first recombinant protein product from a private company, Shantha Biotech, a hepatitis B vaccine, and it has numerous industrial relationships, including a joint venture with Biological E and Amersham Pharmacia to build DNA micro-arrays.

INDIAN COUNCIL OF MEDICAL RESEARCH (ICMR)

Another major institution working in the area of biotechnology is the Indian Council for Medical Research (ICMR) under the Ministry of Health. It is the apex body in the country to promote, coordinate and formulate biomedical and health research. Central Government gives full maintenance grant to the Council, to research in communicable diseases, contraception, maternity and child health, nutrition, non-communicable and basic research.

The total allocation for ICMR from the Central Government (Ministry of Health) was ₹ 1470 million in 2000-01, which was 21 per cent higher over the allocation of the previous year, that is 1999-2000 (₹ 1160 million). The Council is also engaged in research on tribal health, traditional medicine and publication and dissemination of information. In the year 2001 ICMR has launched a major programme in the field of genomics (vector, microbial, human) with the initial allocation of ₹ 510 million. One of the major areas of focus is the disease susceptibility gene identification, especially for, communicable diseases like leprosy, tuberculosis, noncommunicable diseases as rheumatic fever or genetic diseases as thalissimia established four centres for developing molecular medicine at All India Institute of Medical Sciences (AIIMS), New Delhi, Sanjay Gandhi Post Graduate Institute of Medical Sciences (SGPGIMS), Lucknow, Jawaharlal Nehru University (JNU), New Delhi and Post Graduate Institute of Medical Education and Research (PGIMER), Chandigarh. Apart from this, ICMR has also established six biomedical informatics units in different parts of the country. It has proposed an allocation of ₹ 1000 million for the Tenth Five Year Plan.

INDIAN COUNCIL FOR AGRICULTURE RESEARCH (ICAR)

In India, agricultural research is being spearheaded by the Indian Council of Agricultural Research (ICAR) under the Ministry of Agriculture. The Council is engaged in conducting research in the field of agriculture, soil and water conservation, animal husbandry, fisheries, dairying, forestry and also agricultural education. The allocation for ICAR from the Ministry of Agriculture was ₹ 13,990 million in 2000-01, which was ₹ 12,060 million in the previous year. It has several research laboratories all over the country conducting research in biotechnology, besides using traditional breeding techniques for different research projects. ICAR has established a National Research Centre on Plant Biotechnology (NRCPB) at the Indian Agricultural Research Institute (IARI), Pusa, New Delhi, which is fully dedicated to work on plant biotechnology. Annual expenditure over the projects at NRCPB is ₹ 150 million. Apart from this ICAR has also supported two research networks, combining ICAR and even non-ICAR research laboratories to work on crops like rice, cotton, brasica and brinzal. ICAR is also implementing a World Bank supported programme called National Agriculture Technology Programme (NATP) through which huge allocations have been made at different research laboratories for strengthening infrastructure for biotechnological research.

ICAR has also collaborated with DBT and Rockfeller Foundation to jointly launch a National Rice Biotechnology Network (NRBN) in 1988. This project helped in evolving a culture of collaboration among different institutions, which has led to publication of several international papers in established journals. This network put together plant breeders and molecular biologists on the one hand and also provoked interests of private sector in R&D. For instance, Mahyco and Rockfeller Foundation worked together on identifying the relevant genes for saving Indian rice from brown plant hoppers. In this collaborative effort, interaction was not limited to private entities only. It is between private and public sector organisations also. Several universities also have tie ups with private sector organisations. For instance, Tamil Nadu Agricultural University (TNAU) attempted joint trials with Monsanto on weed resistance in soil. TNAU was also partner in monitoring the Bt Cotton field trials of Mahyco-Monsanto alliance. TNAU is also holding in exploratory talks with Rasi seeds (it is also in alliance with Monsanto) to conduct and monitor field trials of Bt crops. Similarly, rice research work at National Research Centre for Plant Biotechnology has attracted business interest of companies like Nath Seeds Ltd. and JK Agri-genetics.

HUMAN RESOURCE DEVELOPMENT AND TRAINING

The National Biotechnology Board had launched an integrated short-term training programme way back in 1984, to cope up with growing demand for highly trained manpower. In the first phase (1984-85), 5 universities were selected for initiating M.Sc./M.Tech programme in this multi-disciplinary area. Subsequently, in 1985-86 and 1986-87, the DBT has added 8 universities/institutions for M.Sc/M.Tech/Post-doctoral teaching programmes. Subsequently, DBT was entrusted with the responsibility of evolving curriculum for biotechnology courses and meet the demand for human resources in the field of biotechnology. In 1986-87 a model system of post-graduate/post-doctoral teaching in biotechnology in 7 universities/institutions was launched Some of the specialised M.Sc. courses in marine and agricultural biotechnology were launched in 1988-89 at 3 universities. In 1992-93, DBT supported a five year Integrated Programme in biochemical engineering and biotechnology in Indian Institute of Technology, Delhi and a post-doctoral programme at Indian Agricultural Research Institute, New Delhi.

Now, DBT is supporting 20 M.Sc. courses in general biotechnology, 4 in agricultural biotechnology, one each in medical and marine biotechnology while couple of diploma courses in molecular and biochemical technology 14. The total intake of students in the various postgraduate courses supported by the DBT in the country is around 550 per year. As a part of restructuring of the post doctoral research and training programme, DBT has scraped the on going programme with different institutions and has given this responsibility to Indian Institute of Science (IISc), Bangalore. This is to ensure competitive attitude and quality output in the life sciences. It is being proposed that IISc would award up to 75 fellowships of two-year duration in different streams of biotechnology.

As part of a wider effort for capacity building in institutes of higher learning, full-fledged departments of biotechnology are being set up. The Indian Institute of Science, Bangalore, Indian Institute of Information Technology and Management, Gwalior; and select Regional Engineering Colleges are setting up departments of biotechnology. The All India Council for Technical Education (AICTE) has already approved B.Tech. programmes in biotechnology in eight engineering colleges and has since been advised to develop a model curriculum for undergraduate programmes. All the new departments will have undergraduate, post-graduate and doctoral programmes. Special funding will be provided for the purpose in the Tenth Five Year Plan. Apart from expanding teaching of biotechnology at higher educational institutions a separate module on biotechnology would also be integrated with the school curriculum. The Department of Biotechnology of Government of India will provide the necessary outline of this module so that the National Council of Education Research and Training (NCERT) and the Boards of School Education would be accordingly advised.

Indian University Grants Commission has come out with a scheme to promote higher centres of learning at one place and assist them as much as possible. In this regard, Delhi based Jawaharlal Nehru University (JNU) has been identified by the UGC as centre for excellence in the areas of genomics, genetics and biotechnology. The University has received funds to the tune of ₹ 300 million and is planning to start a new integrated M.Sc./Ph.D programme in life sciences and made to upgrade equipment and library facilities. The new integrated programme in life sciences will reduce the time taken by a scholar to complete his Ph.D by at least within two years. Now the scholar will not be required to undergo a separate M.Phil programme. JNU is aiming at 10 seats for the integrated course and another 20-25 seats in the School of Life Sciences. The University has so far received 40 proposals for possible projects, which can be pursued by them in these fields in the future. Out of the funds that JNU has finally managed to get from the University Grants

Commission on its selection as the University with potential for excellence, ₹ 100 million have been set aside for upgradation of facilities and equipments. The rest of the ₹ 200 million would help University to subscribe to some of the 8,000 online journals, both in the filed of Science and social sciences. Besides this, JNU also announced recruitment of more researchers. The tie-ups with industry are also likely to grow. For instance, researchers at the Centre for Biotechnology (CBT) at Jawaharlal Nehru University in New Delhi have been working for four years on a recombinant anthrax vaccine and soon would start Phase I clinical trials in collaboration with Panacea Biotech Ltd. (New Delhi).

7

Applications of Biotechnology in Horticulture

The requirement of fruits and vegetables is increasing proportionally with the increasing population in the country. How do we keep horticultural production on par with the burgeoning population? Although conventional plant breeding techniques have made considerable progress in the development of improved varieties, they have not been able to keep pace with the increasing demand for vegetables and fruits in the developing countries. Therefore an immediate need is felt to integrate biotechnology to speed up the crop improvement programmes. Biotechnological tools have revolutionized the entire crop improvement programmes by providing new strains of plants, supply of planting material, more efficient and selective pesticides and improved fertilizers. Many genetically modified fruits and vegetables are already in the market in developed countries. Modern biotechnology encompasses broad areas of biology from utilization of living organisms or substances from those organisms to make or to modify a product, to improve plant or animal or to develop micro-organisms for specific use. It is a new aspect of biological and agricultural science which provides new tools and strategies in the struggle against world's food production problem. The major areas of biotechnology which can be adopted for improvement of horticultural crops are –

TISSUE CULTURE

One of the widest applications of biotechnology has been in the area of tissue culture and micro propagation in particular. It is one of the most widely used techniques for rapid asexual in vitro propagation. This technique is economical in time and space affords greater output and provides disease free and elite propagules. It also facilitates safer and quarantined movements of germplasm across nations. When the traditional methods are unable to meet the demand for propagation material this technique can produce millions of uniformly flowering and yielding plants. Micropropagation of almost all the fruit crops and vegetables is possible now. Production of virus free planting material using meristem culture has been made possible in many horticultural crops. Embryo rescue is another area where plant breeders are able to rescue their crosses which would otherwise abort. Culture of excised embryos of suitable stages of development can circumvent problems encountered in post zygotic incompatibility.

This technique is highly significant in intractable and long duration horticultural species. Many of the dry land legume species have been successfully regenerated from cotyledons, hypocotyls, leaf, ovary, protoplast, petiole root, anthers, etc., Haploid generation through anther/pollen culture is recognized as another important area in crop improvement. It is useful in being rapid and economically feasible. Complete homozygosity of the offspring helps in phenotype selection for quantitative characters and particularly for qualitatively inherited characters making breeding much easier successful isolation, culture and fusion of plant protoplasts has been very useful in transferring cytoplasmic male sterility for obtaining hybrid vigour through mitochondrial recombination and for genetic transformation in plants.

In vitro germplasm conservation is of great significance in providing solutions and alternative approaches to overcoming constrains in management of genetic resources. In crops which are propagated vegetatively and which produce recalcitrant seeds and perennial crops which are highly heterozygous seed storage is not suitable.

In such crops especially, *in vitro* storage is of great practical importance. These techniques have successfully been demonstrated in a number of horticultural crops and there are now various germplasm collection centers. In vitro germplasm also assures the exchange of pest and disease free material and helps in better quarantine.

Plant breeders are continually searching for new genetic variability that is potentially useful in cultivar improvement. A portion of plants regenerated by tissue culture often exhibits phenotypic variation atypical of the original phenotype. Such variation, termed somaclonal variation may be heritable i.e. genetically stable and passed on to the next generation. Alternatively, the variation may be epigenetic and disappear following sexual reproduction. These heritable variation are potentially useful to plant breeders.

GENETIC ENGINEERING OF PLANTS

Using techniques of genetic engineering many useful genes have been introduced into plants and many transgenic plants have been developed in which the foreign DNA has been stably integrated and resulted in the synthesis of appropriate gene product. Transgenic plants have covered about 52.6 m hectares in the Industrial and developing countries upto 2001. Genes for the following traits have been introduced to the crop plants.

- *Herbicide tolerance*: Transgenic plants are developed that are resistant to herbicides allowing farmers to spray crops so as to kill only weeds but not their crops. Many herbicide tolerant plants have been developed in tomato, tobacco, potato, soybean, cotton, corn oilseed rape, petunia, etc. Glyphosate is one of the most potent broad spectrum environment friendly herbicide known, it is marketed under the trade name Round up. Glyphosate kills plants by blocking the action of an enzyme (5-enolpyruvyl shikimate-3-phosphate synthase) (EPSPS) an essential enzyme in the biosynthesis of aromatic amino acids, tyrosine, phenylalanine and tryptophan. Amino acids are building blocks of protein.

Transgenic plants resistant to Glyphosate have been developed by transferring gene of EPSPS that over prodoce this enzyme thus inhibiting the effect of Glyphosate. A number of detoxifying enzymes have been identified in plants as well as in microbes. Some of these include glutahthione-s-transferase or GST in maize and other plants which detoxifies the herbicidebromoxynil and phosphinothricin acetyl transferase (PAT) which detoxifies the herbiside PPT (L-phosphinothricine). Transgenic plants using *bxn* gene from Klebsiella and *bar* gene from Strepotomyces have been obtained in potato, oilseed, sugarbeet, soybean, cotton and corn and are found to be herbicide resistance. These transgenic plants reduce the use of weeding labour, farmers cost and increase yield.

- *Engineering pathogen resistance:* Viruses are the major pests of crop plants which cause considerable yield losses. Many strategies have been applied to control virus infection using coat protein and satellite RNA. Viruses are submicroscopic pockets of nucleic acid (DNA or RNA) enclosed in a protein coat and can multiply within a host cell. Use of viral coat protein as a transgene for producing virus resistant plants is one of the most spectacular successes achieved in plant biotechnology. Coat protein gene from tobacco mosaic virus (TMV) classified as a positive strand RNA virus has been transferred to tobacco, making it nearly resistant against TMV. Using gene for nucelocapsid protein resistance has been introduced in crops like tomato, tobacco, lettuce, groundnut, pepper and in ornaments like *Impatiens, Ageratum* and *Crysnathemum* against tomato spotted wilt virus. Use of satellite RNA (SATRNA) makes many transgenic plants resistant toCucumber Mosaic Virus (CMV). Transgenic resistant plants have also been developed against alfalfa mosaic virus, potato virus X, Rice tungro virus, tobacco rattle virus and Papaya ring spot virus.

  During the last decade many resistance genes whose products are involved in recognizing the invading pathogens have been identified and cloned. A number of signaling pathways which follow the pathogen infection have been dissected. Many of the antifungal compounds synthesized by plants which combat fungal infections have been identified. The major strategies for developing fungal resistance have been production of transgenic plants with antifungal molecules like proteins and toxins, and generation of hypersensitive response through R genes or by manipulating genes of SAR pathway. A chitinase gene from bean plants in tobacco and *Brassica napus* showed enhanced resistance to *Rhizoctonia solani.* In another case chitinase gene obtained from *Serratia marcescens*(soil bacterium) is introduced in tobacco making it resistant to *Alternaria longipes* which causes brown spot diseases. Acetyl transferase gene is introduced in tobacco making it resistant to*Pseudomonas syringea,* a causal agent of wild fire disease.

- *Stress resistance:* A number of genes responsible for providing resistance against stresses such as to water stress heat, cold, salt, heavy metals and phytohormones have been identified. Studies are also being conducted on metabolites like proteins and betains that have been implicated in stress tolerance. Resistance against chilling was introduced into tobacco plants by introducing gene for glycerol-1-phosphate acyl-transferase enzyme from *Arabidopsis.* Many plants respond to drought stress by synthesizing a group of sugar derivatives called polyols (Mannitol, Sorbitol and Sion) . Plants that have more polyols are more resistant to stress. Using a bacterial gene capable of synthesizing mannitols it is possible to raise the level of mannitol very high making plants resistant to drought.
- *Fruit Quality:* Tomatoes which ripen slowly are helpful in transportation process. Transgenic tomato with reduced pectin methyl esterase activity and increased level of soluble solids and higher pH increases processing quality. Tomatoes exhibiting delayed ripening have been produced either by using antisense RNA against enzymes involved in ethylene production (Eg ACC synthase) or by using gene for deaminase which degraded l-aminocyclopropane-l-carboxylic acid (ACC) an immediate precursor of ethylene. This increases the shelf life of tomatoes. These tomatoes can also stay on the plant long giving more time for accumulation of sugars and acids for improving flavour. It is produced at commercial level in European and American countries. Tomatoes with elevated sucrose and reduced starch could also be produced using sucrose phosphate synthase gene. Starch content in potatoes has been increased by 20-40% by using a bacterial ADP glucose pyrophosphorylase gene.
- *Pest resistance*: The insecticidal beta endotoxin gene (bt gene) has been isolated from *Bacillus thuringiensis* the commonly occurring

soil bacteria and transferred to number of plants like cotton, tobacco, tomato, soybean, potato, etc. to make them resistant to attack by insects. These genes produce insecticidal crystal proteins which affect a range of lepidopteran, coleopteran , dipteran insects. These crystals upon ingestion by the insect larva are solubilised in the highly alkaline midgut into individual protoxins which vary from 133 to 136 kDa in molecular weight. Insecticidal crystal protein produced during vegetative growth of the cells (VIP)are also found to be highly effective against insect control. Bt resistant plants are already in the market.

- *Male sterility and Fertility restoration*: This is helpful in hybrid seed production. Transgenic plants with male sterility and fertility restoration genes have become available in *Brassica napus*. It facilitates production of hybrid seed without manual emasculation and controlled pollination as often done in maize. In 1990, Mariani and others from Belgium have successfully used a gene construct having another specific promoter from TA29 gene of tobacco and bacterial coding sequence for a ribonuclease gene from Bacillus Sp. (barnase gene) for production of transgenic plants in *Brassica napus*. Here the translated gene prevented normal pollen development leading to male sterilily.

MOLECULAR DIAGNOSTICS

- *Nucleic acid probes*: It is now possible to detect the plant diseases even before onset of symptoms by using cDNA probes. Probes are nucleic acid sequences of pathogen causing organisms labeled with certain markers. cDNA probes corresponding to specific regions of the pathogens can be generated using standard recombinant DNA technique.
- *Monoclonal antibodies (McAb)*: Immunochemical techniques are extremely useful for the rapid and accurate routine detection of plant pathogens and ultimately the diagnosis of plant disease and their relatedness, The introduction of hybridoma technology has provided methods for the production of homologous and biochemically defined immunological reagents of identical specificity which are produced by a single cell line and are directed against a unique epitope of the immunizing antigen. The great potential of McAbs in phytopathological diagnostics is essential because of homogeneous antibody preparations with defined activity and specificity can be produced in large quantities over long periods. Even though hybridoma technology is a laborious and expensive enterprise compared to standard immunization procedures it is going to be widely used for large scale diagnosis.

MOLECULAR MARKERS

The possibilities of using gene tags of molecular makers for selecting agronomic traits has made the job of breeder easier. It has been possible to score the plants for different traits or disease resistance at the seedling stage itself. The use of RFLP (Restriction Fragment Length polymorphism), RAPD (Random Amplified Polymorphic DNA) , AFLP (Amplified Fragment Length Polymorphism) and isozyme markers in plant breeding are numerous. RFLPs are advantageous over morphological and isozyme markers primarily because their number is limited only by genome size and they are not environmentally or developmentally influenced. Molecular maps now exist for a number of crop plants including corn, tomato, potato, rice, lettuce, wheat, Brassica species and barley. RFLPs have wide ranging applications including cultivar finger printing, identification of quantitative trait loci, analysis of genome organization, germplasm introgression and map-based cloning. AFLP is becoming the tool of choice for fingerprinting because of its reproducibility compared to RAPD. Microsatellile or simple sequence repeats (SSRS) markers have also become the choice for a wide range of applications in genotyping, genome mapping and genome analysis.

DEVELOPMENT OF MICROBIAL INOCULAN

Indiscriminate and injudicious use of chemical fertilizers and pesticides for the crop production and control of insect-pests has resulted in pollution of the environment deterioration of soil health and development of resistance by many insects and residue problems. Hence there is a great concern world wide to use safer biofertilizers and biopesticdies in the integrated nutrient management and pest management systems.

Biofertilizers are micro-organisms which fix atmospheric nitrogen or solubilise fixed phosphorus in the soil and make more nutrients available to the plant. Some of the organisms providing major inputs are the biological nitrogen fixing organisms like *Rhizobium, Azotobacter, Azospirillum* and phosphate solubilising organisms like *Bacillus polymyxa, B. magaterium, Pseudomonas striata* and certain fungal species of Aspergillus and *Penicillium.*

The benefits of using micro-organisms as fertilizers are many fold. They are less expensive, nontoxic to plants, do not pollute the ground water nor render the soil acidic and unfit for growth of plants. *Rhizobium* forms nodules on the roots of leguminous plants and help in fixing nitrogen from the atmosphere to ammonium irons which get converted to amino acids in the plant system. Inoculation with this bacteria helps in reducing addition of nitrogenous fertilizers to the soil. *Azospirillum* is also found colonizing inter cellular spaces inside the root system. These bacteria also contribute substantially to the nitrogen requirement of the plant.

Phosphate solubilising bacteria are another group of micro-organisms which solubilise the insoluble phosphorus in the soil and make them readily available to the crop. Mycorrhiza is the symbiotic association of the roots of crop plants with non-pathogenic fungus. They provide nutrients absorbed from deeper layers of soil to the plants. They help the plants in better plant establishment and growth when inoculated. Many fruit crops like papaya, mango, banana, citrus, pomegranate are found to be dependent on this association and are greatly benefited by its inoculation in procuring higher phosphate and other nutrient from the soil. These mycorrhizal associations help the plants in overcoming pathogen attack also. They improve soil characters too.

Genetic modification of microbes: By using DNA recombination technique it has been possible to genetically manipulate different strains of these bacteria suitable to different environmental conditions and to develop strains with traits with capacity for better competitiveness and nodulation. Biopesticides are biological organisms which can be formulated as that of the pesticides for the control of pests.

Biopesticides are gaining importance in agriculture, horticulture and in public heatlh programmes for the control of pests. The advantages of using biopesticides are many. They are specific to target pests and do not harm the non target organisms such as bees, butterflies and are safe to humans and live stocks, they do not disturb the food-chain nor leave behind toxic residues. Some of the microbial pesticides used to control insect pests are *Bacillus thuringiensis* species to control various insect pests. Insecticidal property of these bacteria are due to crystals of insecticidal proteins produced during sporulation. These proteins are stomach poisons and are highly insect specific. Bt toxins could kill plant parasitic nematode too. Number of baculoviruses (BV) nuclear polyhedrosis virus (NPV) is being developed as microbial pesticides both nationally and internationally, A few examples of these are *Heliothis, Spodoptera, Plusia, Agrotis, Trichoplusia,* etc.

Biocontrol agents : These are other microbes which are antagonistic to several pathogenic fungus and are good substitutes to fungicides or insecticide. These are Bacillus sps. *Pseudomonas fluorescens, Trichoderma, Verticillium sp., Streptromyces* sps. etc. These organisms are commercially available.

CULTURAL WEED CONTROL IN HORTICULTURE

The basic principle involved in cultural weed control is to grow a stand of grass that is dense and competitive enough to prevent weed encroachment. Weeds are not the cause of poor turf, but rather the result

Weeds require light, water and nutrients to grow. As turf loses density, light will penetrate the canopy and cause weed seeds at the thatch and soil surface to germinate. Once germinated, weeds can develop rapidly, especially if turf continues to decline.

Weeds probably will receive sufficient light to develop if you can see bare soil or thatch when you look straight down on a stand of grass. Management practices that discourage turf weeds include proper mowing, watering, fertilizing, thatch control and cultivation. The chance for weed encroachment can be reduced by using locally adapted turfgrasses and establishing them during the correct season of the year.

- *Mowing:* To prevent weed germination, mow frequently at the tallest recommended mowing height. Weeds germinate rapidly when turf is scalped by mowing too short or when it is not mowed frequently enough. Both mistakes decrease turf density and cause an open canopy that favours weeds. Experts recommend a range of mowing heights to meet specific turf activities. Lower mowing heights require more frequent mowing. Each mowing should not remove more than one-third of the total leaf height. Annual grassy weeds -- such as crabgrass -- are especially a problem on turfs that lack density as the result of poor mowing.
- *Fertility:* Apply nitrogen at least once a year. Areas that receive chemical weed control but no nitrogen are probably being mismanaged. Even in low- maintenance situations, the first defence against weeds is dense turf. For cool-season grasses, apply nitrogen in the fall. For warm-season grasses, apply it in the summer. Additional nitrogen should be supplied during the growing season to provide adequate growth and density with less attention given to turfgrass colour. On cool-season grasses avoid heavy applications of nitrogen in the spring. That will encourage excessive vertical leaf growth, bringing with it a greater chance for scalped turf and an eventual decrease in turf density. Don't apply nitrogen to dormant warm-season grasses, since it will encourage winter annual weeds, such as chickweed, henbit and speedwell.
- *Establishment:* Weeds can prevent the establishment of grasses at certain times of the year. Spring seeding of cool-season grasses is seldom successful -- because the fertility requirements and frequent watering necessary for turf establishment will also create intense weed pressure from crabgrass and summer annual broadleaf weeds. Consequently, late summer and early fall are preferred for establishment of cool-season grasses. Cool-season turfgrasses that are germinating by early September will not be affected by summer weeds and will usually establish a competitive stand of grass before winter annuals, such as chickweed and henbit, become a problem.
- *Thatch and Cultivation:* These have a dual effect in terms of weed control. Weed seeds below a thatch layer lack the necessary light for germination, so weeds that germinate in the thatch layer usually die because of the poor environment for seedling development in thatch. In this respect, thatch provides some degree of weed suppression. On the other hand, thatch can cause turf decline by harbouring disease-causing pathogens, reducing water infiltration

and tying up pesticides and fertilizers. Power-raking and verticutting are typically used to remove thatch.

- *Cultivation:* Soil compaction from traffic and tight soils will often decrease turf competition and favour some weeds, especially goosegrass and knotweed. Two ways to fix the problem are coring and slicing. These procedures improve the soil surface and the growing environment for turfgrass.

DAMAGED CORN FOR SILAGE

Corn that has been damaged by drought, high temperatures, blight, frost or hail can be salvaged for silage. Quality will not be as high as where corn has reached the dent stage. Feeding value will depend upon the state of development and how it is handled after the damage occurred. Silage from immature corn usually is higher in moisture, does not ferment in the same manner, frequently has a sour odour and is more laxative when fed in large quantities.

Frosted corn has a low carotene content and should be cut as soon as possible. It will dry out quickly and lose leaves. It may be necessary to add water to corn that has frosted and become too dry to pack well. Drought corn also may need added water. When the corn forage is dry, keep the chopper knives sharp and chop as fine as possible.

Immature corn that has been damaged by extremely high temperatures should not be immediately ensiled. Although these plants may never produce an ear, some additional stalk growth and consequently some additional feed may be produced by delaying harvest.

If the plants are harvested for silage soon after they have been extensively damaged by heat, the stalk will have so much moisture that a very low quality silage will result. Nutrients also will decline significantly through seepage.

Corn that has been damaged by leaf diseases such as the southern corn leaf blight often are made into silage. The blight organism is not believed to be toxic to ruminants. It also has been shown that it does not survive the ensiling process. In severe cases, a secondary infection of molds on the damaged areas of the plant may produce a harmful toxin. However, limited research indicates that this is unlikely.

The greatest problem with this type of silage stems from its lack of energy due to reduced grain formation and improper fermentation due to the excessive dryness of the damaged plant. In severe cases where large areas of the corn plant are dead, fermentation problems caused by the lack of plant moisture could arise.

Nitrates

Under certain soil and environmental conditions, the corn plant may store an excessive amount of nitrogen compounds. Nitrate poisoning and drought conditions often are associated, but the condition is difficult to define. The degree of drought and nitrogen availability generally confuses the issue. As the soil moisture level becomes acute, nitrates move towards the soil surface above the corn roots. Some drought-stricken corn may be short of nitrate instead of oversupplied.

If heavy rains occur at this point, excess nitrogen is leached downward and may be taken up by the corn root system. This could cause abnormally high nitrate levels in the plant. Under these conditions, avoid harvesting silage from the stricken fields for a few days.

One measure that may be helpful in reducing the amount of nitrate in the silage is, to allow the corn plant to grow beyond the period of drought damage. If the plant is capable of making some regrowth following a drought, it may be able to use much of its excessive stores of nitrogen for additional regrowth.

The highest concentrations of excessive nitrogen usually are in the lower portion of the stalk. Raising the chopper cutter blade so that the lower 18 or 20 inches of the stalk remain in the field also may reduce the concentration in the silage. Testing for nitrates before silage harvest is frequently unreliable as a feeding guide. Nitrate level in the plant changes rapidly from day to day and usually is reduced about 1/3 in the ensiling process.

If silage is suspected to contain excessive nitrates, it can be detected by test prior to feeding. This analysis should be taken as near to the time when the silage will be fed as practical.

Silo Gases

Lethal gases may occur at any time during silo filling. The greatest danger is 12 to 72 hours after filling, but gas may occur up to 10 days after the last silage is put in the silo.

Silo gases may appear in any ensiled material, grown on any type of soil and under any level of fertilization.

If the gases are present in higher concentrations, two of them, nitrogen dioxide (NO_2) and nitrogen tetroxide (N_2O_4) may be recognized by their irritating odour and colour. Nitrogen dioxide is reddish brown and nitrogen tetroxide is yellow. Nitric oxide (NO) is a colorless gas that may be present in deadly concentrations without being visible. The gases are heavier than air and remain beneath the air mass over the silage. They often will layer on top of the silage just below the upper edge of the top door. They may settle down the chute to accumulate at the bottom of the silo, in unloading chutes, in adjacent feed rooms and may even move into the barn, loading areas or milk houses. They usually leave a yellow stain on silage, wood or other materials they contact. The presence of dead birds and small animals around these areas is one indication that the gases are at dangerous levels.

A few simple rules will prevent tragedy and injury:

- Run the blower 15 to 20 minutes before going into a partly filled silo. Keep the blower running while anyone is inside.
- Stay out of the silo for at least a week or preferably two after the silo is filled.
- If you experience the slightest throat irritation or coughing, get into fresh air quickly. Immediate treatment by a doctor is an absolute must.
- Ventilate the silo room for at least two weeks after filling by opening outside doors and windows to carry away fumes. Removing the chute doors on the silo down to the level of the settled silage will permit natural ventilation where gas tends to be concentrated.
- Keep the doors between the silo room and the barn closed to protect livestock.

Length of Cut

Corn silage should be cut into particles 1/2 to 3/4 inches in length. Particles of this size will pack more firmly in the silo and are more palatable to cattle. Very finely cut silage may be made with a recutter. This will increase the amount of dry matter that can be stored in a silo, but very finely cut silage is less palatable and has resulted in lower butterfat tests when this feed was the primary source of roughage for dairy cattle.

Adding Water to Dry Silage

If silage is too dry, it may be necessary to add water in order to establish airtight conditions. As a rule of thumb, add four gallons of water per ton of silage for each 1 percent desired rise in moisture content. Add this water as the silo is being filled. If water is added after the silo is filled, it tends to seep down the silo walls and does not permeate the silage mass. This may cause leaching of silage nutrients, seepage that may break the air seal and improper fermentation. Frozen silage: Frozen silage is sometimes a problem, especially with trench or bunker silos. Freezing does not impair the keeping quality of silage so long as the silage is not disturbed, but frozen silage may cause digestive disturbances when eaten by cattle. It is best to thaw silage before feeding it.

Silage additives: Top quality silage can be made without the addition of any additives or preservatives. There is no reliable evidence that adding enzymes, yeast cultures, antibiotics or acid-forming bacteria will increase the feeding value of corn silage. The two additives most often used in corn silage are limestone and non-protein nitrogen compounds.

Molasses and grain may be added to corn forage at the time of ensiling. However, this practice only tends to enrich the resulting corn silage as a feed rather than to improve the quality of the corn forage itself. There is also some fermentation loss with the added molasses and grains.

Molds: Molds are common in silage, especially around silo doors and edges of bunker silos. White and gray molds, which are caused by exposing silage to air, are seldom toxic to livestock, but intake is sometimes reduced. A mold known as monascus causes silage to form lumps or small clumps, which are white on the outside but have a reddish centre. Cattle usually will eat silage containing this mold. There have been no reported cases of toxicity.

In the spring as warm temperatures occur, a rapid-forming red mold may appear on the face of the silage between morning and evening feedings. This is commonly known as bakery mold, monilla sitophilia. The mold spores are activated by the heat produced in the silage process and grow quickly when exposed to air. It has not produced any known cases of toxicity.

Silo Capacities and Filling

Tables giving estimated silo capacities often vary due to the crop ensiled, length of cut and effectiveness of distribution and packing at filling time. Most tables are based on silage harvested at 30 to 35 percent dry matter.

Silo capacities actually vary little in terms of dry matter storage, despite differences in moisture content at filling time. An exception to this is material that is 50 percent dry matter or more at the time it is placed in the silo. With this type of material, dry matter capacity may be reduced as much as 10 percent for estimates based on 30 to 35 percent dry matter.

With large-capacity silos and high-speed filling methods, the distribution and packing of silage in silos has been sacrificed. Improper distribution and packing may cause excessive seepage, poor fermentation and losses in storage capacity. Half the capacity of the silo, which is 14 feet in diameter, is in the outside 2 feet. If an overabundance of fluffy, light material is in this outside area, silo capacity may be reduced as much as 20 percent. Most large tower silos should be equipped with a distributor to aid in proper silage distribution and packing. Length of cut of corn should be 1/2 to 3/4 inches to ensure proper placing.

Covering Filled Silos: A loss of nutrients occurs in all silage during the ensiling process. Microorganisms that carry out the fermentation process require energy. This loss depends on the effort made to exclude air at time of filling and to prevent loss of carbon dioxide, which is necessary to arrest respiration of the ensiled plant cells, as well as prevent seepage losses and undesirable fermentation or spoilage due to surface exposure. There is probably a greater range of losses within a particular type or design of silo than occurs in different types of silos where equally good ensiling practices are followed. So, regardless of silo structure, good ensiling practices save feed.

A plastic cover on a trench or bunker silo or a large-diameter tower silo can materially cut feed losses. For best results, the cover must be applied immediately after the last load is packed in the silo. On trench or bunker silos, it is important to mound or crown the forage so that the rain water will drain off the silo. Plastic covers that are not weighted to hold them firmly on the surface of the silage will be only partly effective. Covering the plastic with old tires plus limestone, sawdust or similar material can be of great benefit.

HARVESTING

Corn should be harvested for silage after the ear is well dented but before the leaves turn brown and dry. The quantity and quality of corn silage are at their peak in this stage of development. The ear has accumulated most of its potential feeding value, but there has been little loss from the leaves and stalks. After the dent stage, feeding value of corn stalks and leaves decreases while field losses increase.

Fall temperatures influence the maturity rate of the grain. Maturity usually refers to the time when the ear has accumulated 100 percent of its dry matter production potential. In many years this potential is not achieved because of cool temperatures and cloudy weather. Values listed may be used as a guide to determine when maximum dry matter production has occurred, but variety and weather interactions will exhibit some influence on the result. Ears usually will be well dented somewhere between the 32- to 35-percent moisture stage.

Table. Relationship of Kernel Moisture to Yield Potential

Water in kernels	Yield of grain as percent of maximum
40 %	93.5
38 %	94.8
36 %	96.3
34 %	98.0
26 %	100.0
21 %	98.0

Corn harvested for silage in the milk or dough stage will yield less feed nutrients per acre than if harvested later. Corn also may ferment improperly in the silo if harvested too soon. Corn silage often is made too early because it is believed that feed is lost if undigested corn kernels appear in the manure. This is not true. Digestibility is as high for well-dented kernels as for immature corn grain.

Corn silage that is cut late and has brown and dead leaves and stalks usually will make fair- to good-quality silage, but total production per acre may be sharply reduced. Field losses as high as 30 percent have been found when silage is made late into the fall or early winter. A 10 percent reduction in the amount of dry matter stored in the silo also has been noted with late-cut silage.

BIOTECHNOLOGY, GENETIC ENGINEERING AND THE SEED

"*Bio*technology," deals with *live* organisms. These may be plants or animals, visible to the naked eye, or microorganisms seen only through a microscope. Agricultural biotechnology—our focus here, encompasses various methods used to modify (or create new or modified) plant life. Such life forms may range from simple one-celled algae or complex multi-celled trees. As earlier noted, research scientists in this field use biological processes involving organisms or natural substances to develop or modify products. Their aim is to change the genetic code of living cells or seeds which is carried in the molecule known as DNA (deoxyribonucleic acid). This is the process we call "genetic engineering."

DNA is composed of two alternating chemicals joined together by two other base chemicals. The sequence of the latter determines the instructions that direct cells to make the various proteins that support life. By manipulating this sequence, corporate scientists can enable these cells or seeds to *perform new functions* or to *produce new substances*. This is done by decoding the order or "sequencing" of a gene's building blocks (*i.e.*, its chemical units), and then changing that order. The latter process is known as "splicing". Through this series of steps, scientists can add or remove traits from a plant that are desirable or undesirable. They can make crops resistant to drought or poisonous to pests and weeds. The process of adding or removing traits to or from plants becomes most controversial when it is done across species. That is, foreign genetic material is inserted into the cells of organisms of a different species. This is how "transgenic organisms" are formed. The product may be something as simple as a carrot containing milk proteins or as complex as a genetically manipulated animal and even human forms.

No one doubts that biotechnology's potential for good is immense — in both medicine and agriculture. Progress has been made in finding medicines that can cure heretofore fatal diseases. Methods of "gene therapy" are already in use whereby a "sick" gene is replaced by a healthy one. But in addition to curative ends, experiments may be undertaken for eugenic purposes, *e.g.*, to determine the height, weight, hair colour and other traits of a fetus. Indeed, a popular test in India and elsewhere enables parents to discard unwanted female fetuses until the prized male fetus materializes. A patent pending in the U.S. patent office proposes to genetically change sperm cells so that the resulting altered animals can pass down the new traits to subsequent generations. Such experiments in animal eugenics will undoubtedly pave the way to human eugenics. Hitler's dream of a "superior" race may be realized within a century of his death

The industries most active in these kind of genetic engineering methods are in the biotech, pharmaceutical and chemical fields. The top companies in these industries include, in addition to Monsanto, DuPont, Bayer, Upjohn and others. Once they have identified and patented plant genes (as well as animal), they will be able to control agriculture and the increasingly powerful food processing industry which manufactures much of the food products that we put on our tables, especially in the Western world. But as these giant TNCs expand their operations and with that their influence, in the economies of Third World countries, their agricultural production and food consumption patterns will undoubtedly reflect that influence and power. For they will have acquired patent monopolies of not only the "discovered" genes but also of the genetic engineering methods used to manipulate these genes. Inherent in this kind of experimentation is the familiar dilemma of humankind of "letting the genie out of the bottle." It's difficult to put it back in. This dangerous scientific game with genes will create transgenic forms that will not only reproduce, but in the process of reproducing could be transformed, and having been transformed, could literally move on — by wind, rain, pollen and other natural ways, to other crops or to wild plants. Eventually, these novel organisms could displace the existing forms, homogenizing agricultural life.

It is safe to say that it is in the agricultural field that genetic engineering is most popular. The easily anticipated reason is the profits it promises. Corporate scientists in agribusiness stress the remarkable increase in yield that they claim will end starvation and malnutrition. But as with any new technology—from the wheel to nuclear power—there is a fine line between good and evil. It is that fine line that is stressed by critics who fear that the evil consequences may overwhelm the good. Whether they are pessimists, as some say, or realists, as viewed by others, the end results they foresee do not bode well for humanity and especially for the "wretched of the earth."

Since the bulk of research money comes from agribusiness, the research will focus on foods that sell well in the DC markets. As has already happened in many countries in the South, arable land is increasingly allocated to cash crops that can be exported and earn foreign exchange. Moreover, farmers who used to grow several products are now inclined to grow one or two crops on the same large landholdings. The promise of huge yields from genetically engineered seeds lures even the small-scale farmers who make up the bulk of the farming community in the Third World. But once hooked on such seeds, they are locked into the contract they were required to sign when they bought these seeds.

Cargill "generously" gave free hybrid seeds to farmers in the Philippines. Instead of the increased yields and lower use of chemicals promised by Cargill, quite the opposite happened. The seeds were not suited to the moist climate of the Philippine Islands, and more pesticides, weed killers and fertilizers had to be applied. Worse still, owing to the restrictive terms of the contract signed, they were prevented from using their neighbours' seeds, as traditionally done when a crop failed. Even worse was the contract sold to Indian farmers by Monsanto which demanded a hefty 'technical fee' per acre in addition to the amount paid for the seed. But they were also locked into Monsanto's chemicals as well, and were subject to a fine if they used someone else's chemicals. Activist Vandana Shiva's description of Monsanto's methods as totalitarian is clinched by the provision that allows Monsanto inspectors to visit these farmers (even in the absence of the land owner) in order to determine whether they have violated any of the terms of the contract.

Beneath a façade of official concern about the effect of invading "seed giants" on Indian agriculture, there is a lot of support on the part of Indian agribusiness and allied interests with ties to outside agricultural ventures for opening up India to foreign agribusiness. It is argued that such ties will bring the technology and capital needed to make Indian agriculture more productive than it is now. Mixed signals have come from Seattle regarding the Indian Delegation's policy stance during the doomed Conference. What bears watching is the evolution of the BJP Government's agricultural pronouncements in the aftermath of the Seattle debacle. Thus far, these have not been reassuring.

The status of a proposed law, the Protection of Plant Varieties and Farmers' Rights Bill, remains unclear. It has been reported by the Press Information Bureau of India that this Bill was introduced in the Lok Sabha on December 3rd. But the same agency announced on December 15, 1999 that this Bill was "likely to be introduced shortly in Parliament. " Nevertheless, the draft that has been circulated has raised more questions than it has answered. One critic states that this bill "will end up benefiting the large seed corporations, some very large farmers, and corporate farming agencies." Such a Bill, if enacted, may well signal the nature of relations we can anticipate between India and powerful industrial nations like the United States — the principal architect of the new corporate order. Will India submit to "the inevitable" or will it stand and fight?

KEEP OUT PATENTS ON CONVENTIONAL SEEDS AND ANIMALS

For several years, patents on genetically modified seeds and animals have been granted worldwide. The damaging impacts on farmers, who are deprived of their rights to save their seeds, and on breeders who can no longer use the patented seeds freely for further breeding, are well known.

In Canada and the U.S., for example, the multinational seed company Monsanto has sued many farmers for alleged patent infringements. The same company has also filed court cases against importers of Argentinean soy to Europe. Furthermore, the possibility of patenting seeds has fostered a highly concentrated market structure with only 10 multinational companies controlling about half of the international seed market. Many farmers organisations and NGOs around the world are fighting against these patents. Because genetically modified organisms (GMOs) are still not grown in most countries, or only used in a small number of crops, the negative impacts of these patents are not being felt everywhere.

However, there is an alarming new trend for patents not only to be claimed on GMOs (such as Round-up ready soybeans), but also on conventional plants. For example, patent claims have been made for soy beans with a better oil quality covering parts of the plant genome when used in conventional breeding and technologies to improve conventional breeding (such as marker assisted breeding).

Some of the most threatening examples in this context are patent applications from Syngenta which claim huge parts of the rice genome and its use in breeding of any food crops that have similar genomic information to rice (such as maize and wheat).

The European Patent Office has also granted a patent on aphid resistant composite plants which are based on marker assisted breeding. Other recent patent applications by Monsanto on pigs are also related to normal breeding methods, indicating the increasing danger of agricultural genetic resources becoming monopolised by a few multinationals on a global scale.

Soon the Enlarged Board of Appeal of the European Patent Office will decide on another patent of this kind - for a method of increasing a specific compound in *Brassica* species.

This decision will determine the patentability of conventional seeds in Europe. Whereas patents on conventional plant varieties are normal practice in the U.S., many other countries, especially developing countries, do not grant patents on plants or animals. But as the recent history shows, the standards defined and used at the European, Japanese and U.S. patent offices influence international regulations (the WTO agreement on trade related aspects of intellectual property rights, TRIPS, and the World Intellectual Property Organisation, WIPO). Patent offices all over the world are pushed to adapt their regulations and practices either through the international regulations or by bilateral agreements. India, for example, has just passed a third patent amendment in order to adapt its law to the TRIPS regulations.

This frightening new trend in patent policy will affect many more farmers and breeders, than has been the case with GMO patents. Any remaining farmers rights and breeders' access to plant varieties and animal breeds for breeding purposes, will disappear everywhere. These patents will destroy a system of farmers' rights and breeders' privileges that has been shown to be crucial for the survival of farmers and breeders, for food sovereignty, and for the preservation of bio-diversity in agriculture. The vast majority of farmers in developing countries are small-scale farmers, completely reliant on saving and exchanging their seeds.

In order to secure the continued existence of independent farming, breeding and livestock keeping and hence the food security of future generations, we, the undersigned farmers, researchers, breeders and civil society organisations from all over the world, restate our rejection of any patents on life, and urge policy makers and patent offices to act swiftly to stop any patents being granted on conventionally bred plants and animals and on gene sequences for use with conventional breeding technique, as well as on methods for the conventional breeding of plants and animals. We also urge companies not to apply for any patents of this kind.

Savage Garden: Slack-Potting the Dewy Pine

Over twenty years ago, British nurseryman Adrian Slack originated a rather clever way to successfully grow the often temperamental dewy pine (*Drosophyllum lusitanicum*). Glaringly omitted from my own book, The Savage Garden, I wish to correct that oversight here, for this method, called "Slack-potting"—a term coined by Barry Rice—is an ingenious method of cultivating *Drosophyllum* with long-term ease.

We must remember that dewy pines are not wetland plants like most other carnivores. Instead, they are typically found on sandy, gravel slopes that are dry most of the year, and usually alkaline in pH. The plant is found in widely scattered sites in Portugal, Spain, and Morocco. The climate is warm-temperate and Mediterranean, which means that most of the rainfall occurs in the cool winter months, with very rare frosts. Summers are warm and dry, but plants near the coast can experience early morning fog drip, which may be absorbed by their leaves.

Seed is usually the only way to propagate *Drosophyllum*, and since plants despise root and stem disturbance they are never sold through the mails. Seed is usually produced in late spring or early summer and can germinate readily if very fresh. Older seed benefits from several methods of pretreatment. You can scarify the hard seed coating by rubbing the black, egg-shaped seeds with sandpaper until the whitish interior becomes slightly visible. You can also soak the seed for a day or two in a cup of water in which you have added a tiny bit of powdered gibberellic acid—about 0.5 cm (1/8 inch) collected on the tip of a toothpick works well. I have also scarified the seed and then soaked them in a solution of one-drop Superthrive™ per cup of water.

Slack warns against moving freshly germinated seedlings about, and prefers to sow a few seed directly in their permanent pots, allowing only one plant to grow by removing any others that germinate. This can be a hassle and waste of seed if you have more than just a few to sow, but Slack's warning is legitimate since the first tiny root to appear from a seed will rapidly plunge into the soil medium and can be killed if disturbed. I prefer to sow numerous seed on damp vermiculite (they will germinate on any damp substance, even a sponge) and I check them daily after a week or two.

Now this is where the Slack-potting begins. Take a 10-15 cm (4-6 inch) pot made of unglazed terra-cotta clay. Place a wad of damp, long-fibered sphagnum moss through its drainage hole, totally blocking it. As a soil medium, I use one part each perlite, vermiculite, and horticultural sand to be excellent. Slack includes a portion of John Innes Compost #2 (a humus-based house plant soil), but I find this component to be unnecessary. (It is only available in Britain.) Dampen the mix with purified water and fill the clay pot firmly with it to the brim. Gently place your germinated seed on the centre surface of this, one seedling per pot. Place the pot in a shallow water tray in a sunny place with good air circulation, keeping the soil damp to wet for several weeks as the plant establishes itself.

Let the baby plant grow for several weeks, as Slack recommends, before proceeding with phase two. The plant will grow rapidly and when it has several leaves you are ready to Slack-pot it.

You will need another pot with drainage holes at least two inches larger than the pot containing the plant. Slack recommends another clay pot, but in hotter, drier climates like California, I use plastic. Block the drainage holes with long-fibered sphagnum. Fill the pot with the perlite-vermiculite-sand medium only as deep as to allow the smaller potted plant to sit comfortably on the medium and have its rim at the same level (or just slightly above) as the larger pot. In the gap between the two pots, tightly pack long-fibered sphagnum.

The whole reason for this elaborate double-potting is to keep the soil around the stem and upper roots on the dry side, or just barely damp, since the prostrate stems tend to rot if left on permanently wet soils. Initially you can keep the whole double pot in a shallow tray of water as well as gently watering the plant itself, keeping the soil damp. But in a very few months the roots of your dewy pine will enter the soil of the larger pot through the sphagnum-crocked hole of the smaller. You can then water the plant via the sphagnum-moat that separates the two pots.

Alternatively, you can set the Slack-potted plant in a shallow water tray, and water by the tray system only enough to keep the sphagnum-moat damp. Just enough moisture will permeate the inner clay pot to keep your dewy pine happy, and the deepest roots will find plenty of moisture in the medium of the larger container. This probably recreates the natural situation well, for in the wild dewy pines have roots that are extensive, obtaining moisture from deep underground. Some also believe the plants absorb moisture in summer from condensing early morning fog, but misting the plant is not required. Slack-potting a dewy pine works best for greenhouse cultivation, particularly in humid climates with long, cool winters, such as Slack's own England. It is not quite as necessary if you live in a Mediterranean climate similar to the dewy pine's.

I have seen photographs of huge *Drosophyllum* plants grown in giant clay pots outdoors in Australia. Geoff Wong's prize winning plant photographed in The Savage Garden was a judiciously watered specimen grown in a plastic pot of primarily perlite and a tiny bit of peat, then placed in that lovely glazed ceramic ornamental. I myself grew one outdoors on my deck in northern California, using a plain clay pot. The plant survived months of winter rain, a hard freeze, and baking summer heat. So while not for everyone, for many growers Slack-potting is ideal. At my nursery just this past week we Slack-potted several dewy pines. In my imagination, I visualized Adrian visiting us in his white suit and Panama hat, nodding his approval as we shared a glass of wine.

Experiments with Highland *Nepenthes* Seedlings

One of the most difficult things to describe to someone is how to grow a plant that is restricted in its tolerances. Partly to blame is our human perception of these conditions. What exactly is "bright light"? What is "clean water" and how acidic is "acid"? Highland *Nepenthes* are often one of the more perplexing groups to work with. What seems to work for one person fails when someone else tries to reproduce the cultural environment. *Nepenthes villosa,* for example, is not an easy plant to grow from seed; it has a narrow range of conditions necessary to grow well. Beans are easy to grow, highland *Nepenthes* are not!

In order to get better insight into the needs of several species, I conducted five years worth of experiments aimed at quantifying their needs. Instead of subjective terms, I used tools to measure environmental conditions in all test groups. A light meter, conductivity meter, pH meter, recording thermometers and a high quality humidity meter were employed to put some numbers behind the findings. (Cheap meters are dangerous to rely on, so only meters that could be calibrated and tested were used.) This way, anyone could nearly duplicate the conditions that were successful, and avoid potential disaster by measuring variables before plants get put into a bad environment.

The following *Nepenthes* were used in this study: *N. burbidgeae, N. edwardsiana* (Tambuyukon type), *N. fusca, N. stenophylla, N. tentaculata* (Mt. Tambuyukon type and Mt. Kinabalu type), and *N. villosa.* All were raised from donated seed or very tiny seedlings (cotyledon spread 0.63 cm., 0.25 inch). Dead or struggling plants were considered indicators of exceeding the species tolerance to one or more variables within their environment. The bane and beauty of using seedlings is that they do not take long to indicate improper environments; they die quickly!

Several points deserve mention, before getting into the "meat" of this chapter. First, plants grow in a system, where the elements of culture play off each other. For example, higher light levels require greater nutrient levels, as plants need more nitrogen to eliminate photosynthetic (waste) byproducts. Stresses from lower relative humidity might be offset by lower light or more available moisture at the root zone. Thus, I am not suggesting one rigid method, when explaining a successful technique. My intention is to give benchmarks and warning signs.

Second, this chapter was written as a distillation from five years (1996-2001) worth of records. A full description of these experiments would fill several volumes of this newsletter with tedious detail. My intention is to present a simplified summary of what happened during these experiments. One example of this practical focus was to simplify the complex science of light. The usual grower purchases a readily available light source or uses sunlight. The intensity of the illumination striking the plants is determined by the light source, the distance from the source to the plants, and/or the amount of shading. The simplest way to get a reasonably accurate measurement of intensity is to use a light meter, specifying the unit of measure and the source (very important). For spectral analysis of artificial lights used herein, contact the manufacturer.

ORGANIC HORTICULTURE

Organic horticulture is the science and art of growing fruits, vegetables, flowers, or ornamental plants by following the essential principles of organic agriculture in soil building and conservation, pest management, and heirloom variety preservation.

The Latin words hortus and cultura together form horticulture, classically defined as the culture or growing of garden plants. Horticulture is also sometimes defined simply as "agriculture minus the plough." Instead of the plough, horticulture makes use of human labour and gardener's hand tools, although some small machine tools like rotary tillers are commonly employed now.

GENERAL

Double Digging, Vermicompost, Mulches, cover crops, compost, manures, and mineral supplements are soil-building mainstays that distinguish this type of farming from its commercial counterpart. Through attention to good healthy soil condition, it is expected that insect, fungal, or other problems that sometimes plague plants can be minimized. However, pheromone traps, insecticidal soap sprays, and other pest-control methods available to organic farmers are also sometimes utilized by organic horticulturists.

Horticulture involves five areas of study. These areas are floriculture, landscape horticulture, olericulture and postharvest physiology. All of these can be, and sometimes are, pursued according to the principles of organic cultivation.

Organic horticulture is based on knowledge and techniques gathered over thousands of years. In general terms, organic horticulture involves natural processes, often taking place over extended periods of time, and a sustainable, holistic approach - while chemical-based horticulture focuses on immediate, isolated effects and reductionist strategies.

ORGANIC GARDENING SYSTEMS

There are a number of formal organic gardening and farming systems that prescribe specific techniques. They tend to be more specific than, and fit within, general organic standards. Biodynamic farming is an approach based on the esoteric teachings of Rudolf Steiner. The Japanese farmer and writer

Masanobu Fukuoka invented a no-till system for small-scale grain production that he called Natural Farming. French intensive and biointensive methods and SPIN Farming are all small scale gardening techniques.

A garden is more than just a means of providing food, it is a model of what is possible in a community - everyone could have a garden of some kind and produce healthy, nutritious organic food, a farmers market, a place to pass on gardening experience, and a sharing of bounty, promoting a more sustainable way of living that would encourage their local economy. A simple 4' x 8' raised bed garden based on the principles of bio-intensive planting and square foot gardening uses fewer nutrients and less water, and could keep a family, or community, supplied with an abundance of healthy, nutritious organic greens, while promoting a more sustainable way of living.

Other methods can also be used to supplement an existing garden. Methods such as composting, or vermicomposting. These practices are ways of recycling organic matter into some of the best organic fertilizers and soil conditioner. Vermicompost is especially easy. The byproduct is also an excellent source of nutrients for an organic garden.

SOIL

Managing the soil is very important. If your garden is healthy then insects will not attack the plants. Insects only attack plants that are unhealthy. As an example illustrating the relative health of plants, insects will not feed on plants which have been genetically modified so as to reduce foliar damage; thus, genetically modified plants are healthier than unmodified ones. To keep your garden healthy give it organic matter and humus to survive. The most important thing is to give your garden lots of attention and your energy.

PEST CONTROL APPROACHES

Differing approaches to pest control are equally notable. In chemical horticulture, a specific insecticide may be applied to quickly kill off a particular insect pest. Chemical controls can dramatically reduce pest populations in the short term, yet by unavoidably killing natural control insects and animals, cause an increase in the pest population in the long term, thereby creating an ever increasing problem. Repeated use of insecticides and herbicides also encourages rapid natural selection of resistant insects, plants and other organisms, necessitating increased use, or requiring new, more powerful controls.In contrast, organic horticulture tends to tolerate some pest populations while taking the long view.

Organic pest control requires a thorough understanding of pest life cycles and interactions, and involves the cumulative effect of many techniques, including:

- Allowing for an acceptable level of pest damage
- Encouraging predatory beneficial insects to flourish and eat pests
- Encouraging beneficial microorganisms

- Careful plant selection, choosing disease-resistant varieties
- Planting companion crops that discourage or divert pests
- Using row covers to protect crop plants during pest migration periods
- Rotating crops to different locations from year to year to interrupt pest reproduction cycles
- Using insect traps to monitor and control insect populations

Each of these techniques also provides other benefits, such as soil protection and improvement, fertilization, pollination, water conservation and season extension. These benefits are both complementary and cumulative in overall effect on site health. Organic pest control and biological pest control can be used as part of integrated pest management. However, IPM can include the use of chemical pesticides that are not part of organic or biological techniques.

AGRICULTURAL BENEFITS OF ORGANIC FARMING

Organic agricultural practices are based on a maximum harmonious relationship with nature aiming at the non-destruction of the environment. The developed nations of the world are concerned about the spreading contamination of poisonous chemicals in food, feed, fodder and fibre. Naturally, organic farming system is looked upon as one of the means to remedy these maladies there. However, the major problem in India is the poor productivity of our soils because of the low level content of the organic matter.

The efficiency of the organic inputs in the promotion of productivity depends on the organic contents of the soil. There were many resemblances of organic farming principles in the traditional agriculture of India. But the former gives a more open and verifiable scientific foundation than the latter.

A study conducted in USA on the nutritional values of both organic and conventional foods found that consumption of the former is healthier. Apples, pears, potatoes, corn, wheat and baby foods were analyzed to find out 'bad' elements such as aluminum, cadimum, lead and mercury and also 'good' elements like boron, calcium, iron, magnesium sellenium and zinc. The organic food, in general, had more than 20 per cent less of the bad elements and about 100 per cent more of the good elements.

INCREASED CROP PRODUCTIVITY AND INCOME

Field trials of organic cotton at Nagpur revealed that during the conversion period, cotton yield was low compared to the conventional and integrated crop management. However, the yields of organic cotton started rising from third year. Cotton yields under organic, conventional and the mixed systems were 898, 623 and 710 kg/ha respectively at the end of the fourth year of the cultivation. The yield of soyabean under organic farming was also the highest compared to the other two systems.

The Central Institute for Cotton Research, Nagpur conducted a study of economics of cotton cultivation in Yavatmal district of Maharashta. The cost of cultivation of cotton was lower in the organic farming than in the modern system. The low costs were due to the non-use of fertilizers and chemical insecticides. As a result of the low yields during the conversion period, the net income from the organic farm was lesser than the conventional farm. But the yield under organic method increased progressively equalling it to that of the conventional system by the sixth year. The input costs were low under organic farming and with a 20 per cent of premium prices of output, the net income increased progressively from fourth year under organic fanning. The appreciation of net income from organic cotton cultivation by the sixth year was 80 per cent over the conventional crop.

Results reported from 1050 field demonstration cum trials under the National Project on Development and Use of Biofertilizers in different parts of the country show* an increase of 4 per cent in yield in plantation crops, 7 per cent in fruit crops, 9 per cent in wheat and sugarcane, 10 per cent in millet and vegetable, 11 per cent in fibre, condiments and spice crops, 14 per cent in oilseeds and flowers and 15 per cent in tobacco.

A study of 100 farmers in Himachal Pradesh during a period of 3 years found that the total cost of production of maize and wheat was lower under organic farming and the net income was 2 to 3 times higher. Both productivity and premium prices contributed to the increased profitability. Another study of 100 farmers of organic and conventional methods in five districts of Karnataka indicated that the cost of organic farming was lower by 80 per cent than that of the conventional one.

Low Incidence of Pests

The study of the effectiveness of organic cotton cultivation on pests at the farm of Central Institute for Cotton Research, Nagpur revealed that the mean monthly counts of eggs, larva and adults of American BoUworm were far lesser under organic farming than under the conventional method. Bio-control methods like the neem based pesticides to Ti-ichoderma are available in the country. Indigenous technological products such as Panchagavya which was experimented at the University of Agricultural Sciences, Bangalore found to control effectively wilt disease in tomato.

Employment Opportunities

Organic farming requires more labour input than the conventional farming system. Thus, India which has a very large amount of labour unemployment and under employment will find organic farming an attraction. Moreover, the problem of periodical unemployment will also get mitigated because of the diversification of the crops with their different planting and harvesting schedules resulting in the requirement of a relatively high labour input.

Indirect Benefits

Several indirect benefits from organic farming are available to both the farmers and consumers. While the consumers get healthy foods with better palatability and taste and nutritive values, the farmers are indirectly benefited from healthy soils and farm production environment. Eco-tourism is increasingly becoming popular and organic farms have turned into such favourite spots in countries like Italy. Protection of the ecosystem, flora, fauna and increased biodiversity and the resulting benefits to all human and living things are great advantages of organic farming which are yet to be properly accounted for.

PROGRESS OF ORGANIC FARMING IN INDIA

The first conference of NGOs on organic farming in India was organized by the Association for Propagation of Indigenous Genetic Resources in October 1984 at Wardha. Several other meetings on organic farming were held at different places in the country towards the end of 1980s. Here, mention must be made of the Bordi Conference in Maharashtra, the state which was the focal point for the organic farming movement in India. The Rajasthan College of Agriculture with the support of the state government organized a meeting on organic agriculture in 1992. The United Planters' Association of South India organised two national level conferences on organic farming in 1993 and 1995. ARISE is a major organization in the country engaged in the promotion of organic farming. ARISE was founded in 1995 at a national conference of organic farming held at Auroville. ARISE comprises of a supporting network of regional groups aiming at sustainable environment by protecting bio-diversity and promoting organic agricultural practices. The selection of Auroville for the conference was apt as it housed the Arabindo Ashram and the pioneering work under its auspices on building technology, alternative energy research, wasteland development, afforestation and organic agriculture.

By 1980, three groups of Indians had taken to organic farming. The first one consisted of urban educated technocrats for peripheral interest, which did not last long. Educated farmers consisted of the second group whose farming practices were based on scientific knowledge. The third group practiced organic farming through trial and error. The successful organic farmers in India are those who have access to sufficient natural resources like, water and other organic inputs mostly on their own farms. These farms produce crops like sugarcane, areca, cocoa, coconut, pepper and spices. Many of them have shown that switch over to organic farming do not affect yields and income and more importantly, knowledge/ expertise is available for successful adoption of organic farming in the country.

The International Federation of Organic Agriculture Movements estimates that an area of about 41,000 hectares in India is under organic farming representing about 0.17 per cent of the world organic acreage. It also reveals that the percentage of organic area to the total cultivated area comes to only about 0.03 per cent and the total number of farms comes to about 5,661. But, a comparison of our 41,000 ha to Australia, Argentina, Italy and USA clearly indicates that organic farming in India has to go very far even to catch up with that of the leading nations of the world.

Non Governmental Organizations are spearheading organic farming in India. A report in 2002 indicates that about 14,000 tonnes of organic products have been raised in India. They include tea, coffee, rice, wheat, pulses, fruits, spices and vegetables. India exports organic agricultural produces to European Union, USA, Canada, Saudi Arabia, UAE, Japan, Singapore and Australia, among others.

The International Conference on "Indian Organic Products-Global Markets" at the end of 2002 was the first to be held in India. IFOAM predicts that India and China have great potential to be organic farm produce exporters in the future. An important event in the history of the modern nascent organic farming in India was the unveiling of the National Programme for Organic Production on 8"^ May, 2000 and the subsequent Accreditation and Certification Programme on P' October, 2001. The logo "India Organic" was released on 26'*' July 2002 to support the NPOP.

An important progress towards organic agriculture made by India is the increasing awareness of the ill effects of the modern farming system, which the country adopted about 35 years ago. The threat poised by the conventional food products to the human health and the damage done to the ecology are being viewed seriously. Efforts are made to produce healthy foods and the demand. for them is increasing. The importance of the marketing of the organic products is highlighted for the promotion of organic agriculture. Several individuals and associations have taken to organic farming and organic products are available in the large cities to a very limited extent.

Important Step Towards in Organic Farming

The most important step towards organic farming taken by the government was to draw a regulatory framework. It is true that the initiatives by the government to introduce organic farming by laying down regulations came belatedly as many countries have already done this kind of basic work decades ago. The implementation of NPOP is ensured by the formulation of the National Accreditation Policy and Programme. The regulations make it mandatory that all organic certification bodies should be accredited by an Accreditation Agency. The international certification agencies operating in India even prior to these regulations will also have to get accreditation under the new dispensation.

The regulations lay down the institutional arrangements for implementing the national programme for organic production. The NPOP is administered, monitored and implemented for the benefit of farmers, processors, traders and consumers. It envisages a three tiered organisation under the overall guidance of the Union Government with the Department of Commerce, Ministry of Commerce and Industry as the nodal agency. Policy making and declaration of the standards for organic products, recognition of organic standards of other nations, efforts to get our standards recognized by others and coordination with other arms of the government for the successful management of the organic agriculture are the major functions entrusted to the ministry.

The agencies accredited are the Agricultural and Processed Food Products Export Development Authority, Coffee Board, Tea Board and the Spices Board. The regulations cover exports, imports and the domestic trade of the organic products. But the government regulations are applicable to only the exports. So, an organic farm product can be exported only if it is certified by a certification body accredited for the purpose. The categories of products covered under accreditation are organic crop production, organic animal production, organic processing operations, wild products and forestry.

A national level steering committee is functioning as the apex advisory body for assisting the government to promote organic farming in the country. This body consists of representative's form the Ministries of Agriculture, Food Processing Industries, Forests and Environment, Science and Technology, Rural Development and Commerce.

Organic production requires certification after periodic inspections in order to ensure that all prescribed practices are followed. The inspection and certification are done by the agencies accredited to the Accrediting Agencies dealing with the commodity. Inspection and certification agencies can be government departments, NGOs, trade or consumer or producer organisations. Such agencies should be registered bodies, with managements in position, declaring the persons who shall be held responsible for any miscarriage of certification and having proof of adequate field staff to undertake periodic inspections. The continued accreditation of such bodies is dependent upon their record of fidelity to the principles of organic production. They are authorised to award certificates after due satisfaction that practices conformed to those enunciated by the Accrediting Agency in relation to the item concerned. The charges levied by the certifying agencies are fixed by the Accrediting Agencies.

SEVERAL PROJECTS AND INITIATIVES TO PROMOTE ORGANIC FARMING

Several projects and initiatives to promote organic farming in the country have begun at the behest of individuals and institutions. A project aided by the World Bank to empower the rural communities in the country to grow organic products for exports had come up in 2002. The programme aims at the improvement and promotion of organic production of spices, certification and export of black pepper, white pepper, ginger, turmeric, cardamom, clove, nutmeg and herbals like rosemary, thyme, oregano and parsley. The implementation of the progamme is done by the NGOs, and Idukki and Waynad districts of Kerala, Nilgiri district of Tamil Nadu and Kandhamal district of Orissa are the areas selected for the purpose. Imparting training to both the JNTGOs and the farmers on organic production methods, basic standards required, documentation, inspection and certification is a major objective of the programme. The assistance to NGOs includes among others computer hardware and software especially for market promotion of their produces.

An initiative for the spread of organic farming by various stakeholders in the Indian organic agriculture sector is their coming together to constitute an apex body for providing centralized services and expertise for the increasing number of organic farmers in the country. Initial steps for setting up the Indian Competence Centre for Organic Agriculture has been taken at a meeting held at the National Academy for Agricultural Sciences in New Delhi in 2003 under the joint auspices of INDOCERT, a Kerala based organic certification agency and the Swiss based FiBL. The decision was the outcome of an Indian team's visit to Switzerland in 2003 to study the structure of Swiss organic farming.

The meeting elected a nine member promoter board of directors to facilitate the setting up of ICCOA, which would be registered as a charitable society. The Centre would strive to strengthen and supplement the efforts of the state governments in promoting the development of organic agriculture in the country.

In Haryana, an enterprising farmer who began farming on his 16 acre land in the Sonepat district in 1971 could establish an organic farm on 108 acres, raising vegetables and other crops. It appears that his success is mainly attributable to the efforts made by him to market the products. NGOs functioning in the neighbourhood of Delhi buy the produces like rice, wheat, pulses and vegetables from his farm at a premium price of 30 to 50 per cent. An exporting firm at a premium of 20 to 30 per cent buys his basmathi rice. Almost 70 per cent of his farm production is sold through advance agreements/ contracts. Presently he and his friends are working with the resident associations in Delhi to market their organic products at a premium of about 25 per cent.

In Rajasthan, the Morarka Foundation, established in 1995, promotes sustainable agriculture. It has about 10,000 partners producing vermi-compost and the Foundation is said to be the single largest producer of this organic input in Asia. It encourages the production of bio-pesticides and supports procurement of certified organic products. It has set up a joint venture to promote agri-biotechnology parks and a model park of such a nature has come up in Jaipur on 20 acres of land.

The campaign launched by the Foundation in the Sheldiawati region of the state in favour of organic farming has resulted in reduction of cultivation costs and improvement in the quality of the produces. Small and marginal farmers in the districts of Sikar and Jhunjhunu, who had been complaining of degradation of their agricultural land and declining productivity, were benefited. The efforts of the Foundation to popularize the use of vermi-compost in place of the chemical fertilizer by creating awareness and imparting training to the farmers have been successful. The Foundation too has a large facility to produce vermi-compost. Application of vermi-compost reduces irrigation, increases the flavour of the products and results in a decline in the damage to the crops by insects. The Foundation is presently engaged in developing techniques to enrich vermi-compost through micro-organisms to make it suitable for location and crop specific application. The promotion of organic farming is encouraged through natural resource management based on ecology protection and sustainable agricultural methods.

In Kerala, the POABS Organic Estate at Nelliyampathy is engaged in the organic cultivation of several agricultural produces. Research on organic farming methods including manures and pesticides is also undertaken there. Liquid manures, mixtures of slurry, cow urine and some herbal preparations are also made for use in the estate.

Several people have taken to organic farming in the state fully convinced of its beneficial effects on man and nature. A farmer who took organic farming as a mission in the district of Kottayam was an example. His estate grew only green crops, from vegetables to rubber. When the rubber plantation was raised on organic methods, the yields were low during the initial years. On the application of the organic inputs, they increased and after three years the yields were on par with the conventional rubber trees.

An initiative to make the state of Kerala fully organic has begun with the formulation of a draft policy in 2003. A workshop on Organic Sustainability of Kerala - A Global Model was organised jointly by the state agriculture department and the Confederation of Indian Industry to deliberate on the organic farming in Kerala. The government policy will focus on conversion of land, produces and budgetary support. Assistance will also be available for certification and inputs, promoting the local certifying bodies, development of agronomic practices, extension support and training to farmers. Creation of consumer awareness, quality considerations, and emphasis on income rather than on yield will also be part of the policy. A campaign to promote organic farming on the lines of the programme on Literate Kerala is proposed to be launched. Organic farming, it is suggested, can be promoted among the educated youth who are presently averse to conventional agriculture.

The Peerumedu Development Society is a cooperative society in the Idukki district, which promotes organic farming for the last 12 years. Pepper, cardamom, coffee, tea, nutmeg, clove, etc. are the crops raised by about 20000 farmers of the PDS who are encouraged to adopt organic farming. PDS collects the products from its member farmers and exports too. It has also started an organic tea processing factory in district.

A view currently gaining ground in the state is that organic farming has almost become necessary for the Kerala farmer, reeling under the onslaught of the highly sensitive international markets for agriculture products. The close linking of solid waste management with bio-manure production is opening up new avenues in city planning in Kerala. The State has short listed three organizations for conducting solid waste management programmmes. The possibilities of making fuel pellets out of the rejected wastes are also being explored. 'Jaivam' or organic is the catchword now gaining popularity in Kerala. The tide seems to be turning in favour of bio-manure, bio-pest control and bio-disease control in all forms of agriculture. The state controlled Vegetable and Fruits Promotion Council, Keralam is also taking slow steps in promoting the 'bio' trend as against the overriding importance given to the chemical methods of farming.

An Association of Agriculturists Functioning in the State

'Infam' is an association of agriculturists functioning in the state. As a part of its efforts to promote organic farming, it is planning to launch a producers' company for marketing of organic produces with a view to assist the growing trend of organic farming in the state. In the initial stage, the functioning of the company will be confined to the Wayanad district. The company will have about 3200 farmers of the district as members who have decided to switch over to organic mode of cultivation. Steps are also being taken to forge collaboration with the Indian Institute of Rural Management, Anand. The proposed tie up with IRMA will be for technical help and also for marketing of the products. The Infam has conducted about 300 training programmes in Wayanad district, which had motivated the farmers to adopt organic farming methods of cultivation.

The MS Swaminathan Research Foundation, Chennai that has been operating in the area, assists the movement. Wayanad district is on the verge of an environmental calamity due to the excessive use of chemical fertilizers and pesticides that depleted the soil beyond redemption. The top soil is fast losing its fertility and is becoming barren in many areas. About 2100 banana farmers used tonnes of Furidan, an insecticide annually in the district alone. The increased use of chemicals was also posing a health hazard. A survey conducted by the Infam among the school children in Thavingal Panchayath recently revealed that 250 of them were cancer patients. The cancer disease in the region is alarmingly on the rise. Karnataka has finalized the policy on organic farming and thus has become the first state in the country to adopt it. It encourages the farmers to adopt organic farming and gradually give up the use of chemical fertilizers and pesticides. The government has earmarked

₹ 20 crores in the state budget of 2004-05 for the purpose. The organic farming movement in the state has reached a stage in which the creation of markets for the products has become essential.

Association for Promotion of Organic Farming at Bangalore consists of 150 voluntary members. It has been promoting organic farming by creating awareness, inspecting and giving organic certification to the farms. Organic farm production in the state is not very significant.

However, some NGOs and individuals are buying organic produces although the movement to convert the land to organic farming has not been successful. There is no organised market for organic produces and if not sold out in the right time, they are left with the conventional produce at the same prices. At the same time, while the farmers are interested in the organic cultivation the consumers are willing to pay for such products. But the link between the two simply does not exist. Consumers demand consistent supplies of good quality organic products and the farmers want an assured market for such produces.

APOP is making the farmers aware of the organic farming methods, particularly about the preparation of the soil for the purpose. The principles behind the healthy soil through the use of the on-farm inputs and compost making are explained to them. Purushotthom Rao who owned 10 acres of land in the Shimoga district took up organic farming in 1989. He raised coconut and paddy along with several other crops in place of arecanut which was destroyed due to strong winds. The yield of coconut he obtained was 80 per cent more than that of his conventional farming friends. The cost of paddy cultivation incurred was 80 per cent lesser than his said friends.

An enterprising farmer practicing organic farming could export several products to the European market. His firm sent 5000 kg of black pepper, 1000 kg of nutmeg, 500 kg of mace, 4000 kg of white pepper, 1000 kg of clove, 300 kg of vanilla, 2000 kg of turmeric, 2000 kg of ginger, 5000 kg of white hibiscus, 80000 kg of henna and 500 kg of cardamom. His firm is certified by a European agency as organic since 1994.

The state of Manipur in the north-east> of India has decided to encourage organic agriculture. The north-east region with its unique characteristics and agricultural practices can be a potential area for the introduction of organic farming. Manipur is aiming at the encouragement of the eco-friendly farm practices and pollution free industry in the state. Dependence of the agro-chemicals will be avoided and factories and industries causing harm to plants and humans will be discouraged. The emphasis on protection of environment is in perfect harmony with the identification of the entire north-east as the mega diversity spot containing rich reserves of flora and fauna. Given the geographical location and agro-climatic zones, the state gives importance to horticulture and agro-industries. The National Horticulture Board has identified 2.77 lakh hectares for the exclusive development of horticulture in the state. It has been decided to encourage organic farming in this development area. The state has proposed to the setting up of an exclusive export zone for organic products in collaboration with the Agriculture Processed Food Export Development Authority. Tamil Nadu plans to encourage organic farming in horticulture and plantation crops to increase the income of the farmers of the state. The state government also wants to promote organic cultivation of fruits, vegetables and tea. The Tamil Nadu Agricultural University has established a model organic farm on a 2.5 hectare area in the campus.

The National Bank for Agriculture and Rural Development has decided to promote organic cultivation of horticultural crops in the mango producing areas of southern Tamil Nadu districts. Theni and Dindigul are major mango growing areas and the fact that the state government has decided to promote an agri export zone for mangoes in these districts will quicken the efforts for organic farming. The aim is to take the benefits of the premium prices for organic products in the developed countries through exports of mangoes. NABARD has funded a farmer in Theni district to the tune of ₹ 3.5 crores for organic mango cultivation on a 252 acre area. Identification of land for organic cultivation was a difficult process and it took one year for the farmer to do so and another three years to convince the banks. Search for the ideal planting materials took several months and he incurred an annual cost of ₹ 46000 for certification.

A seminar organised at Shri Murugappa Chettiar Research Centre at Chennai focused on the need for natural cultivation of food products. Application of farm manure, biological pest control, companion planting, inter cropping, introduction of beneficial microbes and fungies were advocated for natural food farming. The importance of earthworms and the harm caused to them by chemical farming were highlighted at the discussions. In the history of organic farming in Tamil Nadu, the efforts made in 1970s to revive the traditional methods of agriculture at Chennai were praiseworthy. A study held by MCRC in the city revealed that 58 per cent of the people were aware of the benefits of organic farming and 82 per cent were ready to buy organic foods if the prices were competitive.

A few farmers in Puliangudi village in Tirunelveli district have successfully adopted a package of eco-friendly technologies in the paddy cultivation. It is found that these practices show good results on the indigenous rice varieties. The cost of cultivation has substantially been reduced and the farmers receive a premium for the organic rice. The cost of cultivation worked out comes to about ₹ 8,750 per hectare. The price of rice obtained was ₹ 30 per kg. The cultivation turned out to be rewarding economically besides being environmentally acceptable. Several farmers are coming forward to practice the method of organic farming after witnessing the results. Another farmer of the same village revealed that he got about 9,250 kg of paddy per ha and no plant protection was done after adoption of the new method. His cost of cultivation worked out was about ₹ 12,500 per ha. In Goa, a co-operative institution, the Adarsha Krishi Sahakari Kharedi Vikri Prakriya Sanstha, pioneered organic farming of cashew and coconut crops. The organisation has 1000 members and is led by its Chairman under whose leadership the organic cultivation is launched. The Cooperative Society has been receiving export orders from traders from abroad for supply of organic cashew nuts. The Society has helped more than 450 cashew growers to convert to organic farming by imparting training and other supports. It also has a processing facility for about 100 tonnes of cashew nuts during the season. The price paid for a kg of organic raw nuts is ₹ 35.50. The processed nuts are taken to Mumbai and Delhi by the Society for sale.

The state of Madhya Pradesh has gone for organic soya cultivation. Plans are also ready to convert maize, wheat and pulse cultivation to organic on the principle of crop rotation. Organic soyabean is cultivated in the Malwa region of the state in the districts of Khargone and Dhar on an area of 8,700 hectares under contract farming. Farmers who have not used chemical fertilizers and pesticides during the last two years on the land offered for organic farming are selected.

This condition was necessary to restore the soil health to make the land suitable for organic cultivation. Certification of the farm has been done by SKAL, an international agency accredited to APEDA. On getting the certificate, the farm can use the logo 'India Organic' on soyabean and its products like soya oil and deoiled cake. The production is meant for both the national and international markets. It is said that the conventional yellow soyabean fetches ₹ 12,000 per tonne in the local market, while the organic soyabean rules at ₹ 34,000 per tonne in Delhi. However, a price premium of 50 per cent is the least expected for the organic soya. This will be financially beneficial to the farmers, even though the productivity of conventional soyabean is put at 800-900 kg, which is higher than 500-600 kg of the organic soya. The genetically modified soya crops dominate the international market for soyabean and its products. However, the demand for organic soya is high ensuring a good premium for exports from the country.

MARKETING PROBLEMS OF ORGANIC INPUTS

Bio-fertilizers and bio-pesticides are yet to become popular in the country. There is a lack of marketing and distribution network for them because the retailers are not interested to deal in these products, as the demand is low. The erratic supplies and the low level of awareness of the cultivators also add to the problem. Higher margins of profit for chemical fertilizers and pesticides for retailing, heavy advertisement campaigns by the manufacturers and dealers are other major problems affecting the markets for organic inputs in India. Promotion of organic agriculture both for export and domestic consumption, the requirements of food security for millions of the poor, national self-sufficiency in food production, product and input supplies, etc. are vital issues which will have to be dealt with in an appropriate agriculture policy of India. These are serious issues the solution for which hard and consistent efforts along with a national consensus will be essential to go forward. Formulation of an appropriate agriculture policy taking care of these complexities is essential to promote organic agriculture in a big way.

The developing countries like India have to design a plethora of national and regional standards in attune with those of the developed countries. The adoption and maintenance of such a regulatory framework and its implementation will be costly. The cost of certification, a major component of which is the periodical inspections carried out by the certifying agencies, which have freedom to fix the timings, type and number of such inspections appears to be burdensome for the small and marginal farmers. Of course, the fees charged by the international agencies working in India before the NPOP were prohibitive and that was a reason for the weak response to organic agriculture even among the large farms in the country. No financial support as being provided in advanced countries like Germany is available in India. Supports for the marketing of the organic products are also not forthcoming neither from the State nor from the Union governments. Even the financial assistance extended to the conventional farming methods are absent for the promotion of organic farming.

In many cases the farmers experience some loss in yields on discarding synthetic inputs on conversion of their farming method from conventional to organic. Restoration of full biological activity in terms of growth of beneficial insect populations, nitrogen fixation from legumes, pest suppression and fertility problems will take some time and the reduction in the yield rates is the result in the interregnum. It may also be possible that it will take years to make organic production possible on the farm. Small and marginal farmers cannot take the risk of low jaelds for the initial 2-3 years on the conversion to organic farming.

There are no plans to compensate them during the gestation period. The price premiums on the organic products will not be much of help, as they will disappear once significant quantities of organic farm products are made available. The demand for organic products is high in the advanced countries of the west like USA, European Union and Japan. It is reported that the US consumers are ready to pay a premium price of 60 to 100 per cent for the organic products. The upper classes in India are also following this trend as elsewhere. The market survey done by the International Trade Centre during 2000 indicates that the demand for organic products is growing rapidly in many of the world markets while the supply is unable to match it.

India is known in the world organic market as a tea supplier and there is a good potential to export coffee, vegetables, sugar, herbs, spices and vanilla. In spite of the several initiatives to produce and export organic produces from the country, the aggregate production for export came to only about 14000 tonnes. This also includes the production of organic spices in about 1000 ha under certification. Some export houses like Good Value Marketing Ltd and Burmah Trading Corporation are also engaged in exporting of organic fruits, vegetables and coffee from India. The country could export almost 85 per cent of the production indicating that demand is not a constraint in the international markets for organic products.

Hybrid seeds are designed to respond to fertilizers and chemicals. The seed, fertilizer and pesticide industry as also the importers of these inputs to the country have a stake in the conventional farming. Their opposition to organic farming stems from these interests. The need for fixing standards and quality parameters for bio-fertilizers and biomanures has arisen with the increasing popularity of organic farming in the country. There are a very large number of brands of organic manures, claiming the high levels of natural nutrients and essential elements. But most farmers are not aware of the pitfalls of using the commercially available biomanure products. While the concept of organic farming itself lays great stress on the manures produced on the farm and the farmers' household, many of the branded products available in the market may not be really organic. lements of chemicals slipping into the manures through faulty production methods could make the product not certifiable as organic. The process of composting which is a major activity to be carefully done is achieved usually by one of the two methods, vermi-composting or microbe composting.

While the former is ideal for segregated waste material without foreign matter, microbe composting is suitable for large scale management of solid wastes, especially in cities and metres. Even though the farmers are using manure produced by different methods, proper parameters for biomanure are yet to be finalized. Most farmers are still unaware of the difference between biomanure and bio-fertilizer, it is point out. While biomanure contains organic matter, which improves the soil quality, bio-fertilizers are nutritional additives separated from the organic material, which could be added to the soil, much like taking vitamin pills. Bio-fertilizers do nothing to enhance soil quality while the loss of soil quality has been the major problem faced by farmers these days.

An understanding of the real costs of erosion of soil and human health, the loss of welfare of both humans and other living things and the computation of these costs are necessary to evaluate the benefits of organic farming. These costs will have to be integrated to a plan for the implementation of organic agriculture. A recent study shows the inappropriateness of the cost and return accounting methods adopted to find out the economics of the organic farming. An economic evaluation of the bad effects of inorganic agriculture and their internalization through environmental taxes is proposed for a market based approach to promote organic farming in India.

Political Interventions

Agriculture in India is subject to political interventions with the objectives of dispensing favours for electoral benefits. Subsidies and other supports from both the Central and state governments, government controlled prices of inputs like chemical fertilizers, the public sector units' dominant role in the production of fertilizers, government support/floor prices for many agricultural products, supply of inputs like power and water either free of cost or at a subsidized rate, etc. are the tools often used to achieve political objectives. Any movement for the promotion of organic farming in India will have to counter opposition from the sections who benefit from such policies in the conventional farming system. The political system in a democracy like India is likely to evade the formulation of policies, which affect the interests of the voting blocks unless there are more powerful counter forces demanding changes. In the absence of alternative employment opportunities and other considerations, the organized workforce particularly in the public sector fertilizer, pesticide and seed industries is also likely to oppose moves on the part of the government to promote organic farming on a large scale.

VEGAN ORGANIC GARDENING

Vegan organic gardening and farming is the organic cultivation and production of food crops and other crops with a minimal amount of exploitation or harm to any animal. Vegan and vegan-organic farmers use no animal products or by-products, such as bloodmeal, fish products, bone meal, feces, or other animal-origin matter, because they view the production of these materials as either harming animals directly, or as being associated with the exploitation and consequent suffering of animals. Some of these materials are by-products of animal husbandry, created during the process of cultivating animals for the production of meat, milk, skins, furs, entertainment, labour, or companionship; the sale of by-products decreases expenses and increases profit for those engaged in animal husbandry, and therefore helps support the animal husbandry industry, an outcome most vegans find unacceptable.

Soil fertility is maintained by the use of green manures and composted vegetable matter and minerals. Some vegan gardeners supplement this with human urine and 'humanure' produced from compost toilets; others avoid the potential health risks of using human waste. Such wastes may technically be considered 'animal products', however many vegan organic growers do not consider their usage unacceptable as there is unlikely to have been exploitation associated with their production.

VEGAN ORGANIC GARDENING

Many vegan organic gardeners prepare soil for cultivation using the same time-honored method used by conventional and organic gardeners of breaking up the soil with hand tools and power tools and allowing the weeds to decompose. Shallow cultivation is also becoming popular among such gardeners: shallowly turning the soil's surface using one of numerous surface cultivating tools that are now available in the market place, including the popular Coleman surface-hoes developed by Eliot Coleman. Shallow tilling disturbs the soil less than deep turning and, combined with avoidance of soil compaction as described by O'Brien helps the soil maintain ecological balance. This minimizes the opportunities for soil diseases, plant diseases, and insect pests to become abundant.

O'BRIEN VEGANIC GARDENING METHOD

The Kenneth Dalziel O'Brien Veganic Gardening Method is a distinct system that was developed by Rosa Dalziell O'Brien, Kenneth Dalziel O'Brien, and May E Bruce, although the term was originally coined by Geoffrey Rudd as a contraction of 'vegetable organic' in order to "denote a clear distinction between conventional chemical based systems and organic ones based on animal manures". The O'Brien system's principal argument is that animal manures are harmful to soil health rather than that their use involves exploitation of and cruelty to animals.

The Kenneth Dalziel O'Brien system employs very specific techniques including the addition of straw and other vegetable wastes to the soil in order to maintain soil fertility. Gardeners following the Dalziel O'Brien system use soil-covering mulches, and employ non-compacting surface cultivation techniques using any short-handled, wide-bladed, hand hoe. They kneel when surface cultivating, placing a board under their knees to spread out the pressure, and prevent soil compaction. Kenneth Dalziel O'Brien published a description of his system in Veganic Gardening, the Alternative System for Healthier Crops. From Veganic Gardening by Kenneth Dalziel O'Brien: "the veganic method of clearing heavily infested land is to take advantage of a plant's tendencies to move its roots nearer to the soil's surface when it is deprived of light. To make use of this principle, aided by a decaying process of the top growth of weeds, etc., it is necessary to subject such growth to heat and moisture in order to speed up the decay, and this is done by applying lime, then a heavy straw cover, and then the herbal compost activator.... The following are required: Sufficient new straw to cover an area to be leared to a depth of 3 to 4 inches..."

Also part of the O'Brien method is minimal disturbance of the soil by tilling, use of compost, mulch, cover crops, and green manures, use of permanent raised beds and permanent hard-packed paths between them, laying out beds from north to south, putting plants in double rows or more so that not every row has a path on both sides.

ORGANIC AGRICULTURE AND YIELD

Yields relative to comparable conventional systems are directly related to the intensity of farming of the prevailing conventional systems. This is not only the case for comparison between regions, but also between crops within a region, and for individual crops over time8.

An over-simplification of the impact of conversion to organic agriculture on yield indicates that:

- In intensive farming systems, organic agriculture decreases yield; the range depends on the intensity of external input use before conversion9.
- In the so-called green revolution areas, conversion to organic agriculture usually leads to almost identical yields.
- In traditional rain-fed agriculture, organic agriculture has shown the potential to increase yields.

A number of studies have shown that under drought conditions, crops in organic agriculture systems produce significantly higher yields than comparable conventional agricultural crops, often out-yielding conventional crops10 by 7–90%. Others have shown that organic systems have less long-term yield variability. A survey of 208 projects in developing tropical countries in which contemporary organic practices were introduced, showed average yield increases of 5–10% in irrigated crops and 50–100% in rainfed crops. The so-called organic transition effect, in which a yield decline in the first 1–4 years of transition to organic agriculture, followed by a yield increase when soils have developed adequate biological activity has not been borne out in some reviews of yield comparison studies. Trials conducted on organic cotton at Nagpur indicated that after the third year, the organic plot, which did not receive fertilizers and insecticides, produced as much cotton as that cultivated with them. Similarly, studies conducted in Punjab clearly indicated that organic farming gave higher or equal yields of different cropping systems compared to chemical farming after an initial period of three years.

CONSERVATION TILLAGE AND ORGANIC FARMING

Organic agriculture is often characterized as addicted to maximum tillage — with growers using every opportunity to lay the land bare with shovel, plow, or rototiller. This image has been magnified through the popularity of small-scale organic systems like the French Intensive and Biointensive Mini Farming models that espouse double- and triple-digging to create deep rooting beds. While appropriate to such intensive systems, this degree of cultivation is not characteristic of organic agriculture in general. It may surprise some to learn that a large number of organic producers are not only interested in conservation tillage, they have adopted it. This will be a surprise because many believe that conservation tillage always requires herbicides.

The interest in conservation tillage among organic producers in the Cornbelt was well documented in the mid-1970s by Washington University researchers. They noted that the vast majority of organic farmers participating in their studies had abandoned the moldboard plow for chisel plows. Plowing with a chisel implement is a form of *mulch tillage,* in which residues are mixed in the upper layers of the soil, and a significant percentage remains on the soil surface to reduce erosion. Furthermore, a notable number of organic farmers had gone further to adopt ridge-tillage, a system with even greater potential to reduce erosion. It was especially interesting to note that the use of these conservation technologies was almost nil among neighboring conventional farms at the time. Organic growers were actually pioneers of conservation tillage in their communities.

Among the more well-known of these pioneers were Dick and Sharon Thompson of Boone, Iowa. Their experiences with ridge-tillage and sustainable agriculture became the focus of a series of publications titled *Nature's Ag School.* These were published by the Regenerative Agriculture Association — the forerunner to the Rodale Institute. They are now, unfortunately, out of print.

Research continues to open up new possibilities in conservation tillage for organic farms. New strategies for mechanically killing winter cover crops and planting or transplanting into the residue without tillage are being explored by a number of USDA, land-grant, and farmer researchers. Notable among these is the work being done by Abdul-Baki and Teasdale at the USDA in Beltsville, Maryland — transplanting tomato and broccoli crops into mechanically killed hairy vetch and forage soybeans. There are also the well-publicized efforts of Pennsylvania farmer Steve Groff, whose no-till system centres on the use of a rolling stalk chopper to kill cover crops prior to planting. Systems like Groff's and Abdul-Baki's are of particular interest because close to 100% of crop residue remains on the soil surface - providing all the soil conservation and cultural benefits of a thick organic mulch.

After blind cultivation, subsequent weed control operations in larger-scale systems can make use of advances in tillage equipment such as rolling cultivators, finger weeders, and torsion weeders that allow tilling close to the plant row. Smaller-scale operations often use wheel hoes, stirrup hoes, and other less capital-intensive hardware. Determining the amount, the timing, and the kind of tillage to be done can be a balancing act for the organic grower, but experience and observation over time lead to proficiency.

There are downsides to tillage, however, and most organic growers are well aware of them. The most obvious of these is the dollar cost; organic farmers are as concerned as their conventional counterparts about costs of production and strive to minimize expensive field operations. There is also a cost to the soil and environment. Every tillage operation aerates the soil and speeds the decomposition of the organic fraction. While this may provide a boost to the current crop, it can be overdone and "burn up" the humus reserves in the soil. Excessive tillage can also be directly destructive to earthworms and their tunneling, reducing their benefits to the land. There is also the danger of compaction, even when field operations are well timed.

PRACTICE OF MULCHING

Mulching is a practice often used by organic growers. Traditionally, it entails the spreading of large amounts of organic materials — straw, old hay, wood chips, etc. — over otherwise bare soil between and among crop plants. Organic mulches regulate soil moisture and temperature, suppress weeds, and provide organic matter to the soil. Mulching is most appropriate to small, intensive operations with high-value annual or fruit crops.

A few systems of no-till organic gardening have evolved from the concept of deep, *permanent* mulching. Among these are the well-known Ruth Stout method and *Synergistic Agriculture* — a raised bed system developed by Emilia Hazelip, who adapted concepts from *Permaculture* and the ideas of Masanobu Fukuoka. Mark Cain and Michael Crane, co-owners of Dripping Springs Gardens — an intensive market gardening operation in Arkansas — have adapted Emilia's system to their farm with considerable success.

Plastic mulch, as long as it is removed at end of growing or harvest season, is also permitted in certified organic production. Its use allows larger acreage to be brought more easily under herbicide-free management, though there are serious issues to be addressed.

High-Input Organic Agriculture

At the beginning of this publication, organic farming was described as a system that uses a minimum of off-farm inputs. While that describes most of organic agriculture as it is currently practiced in the U.S., certified organic farming can also entail much greater reliance on off-farm inputs.

Intensive annual strawberry and vegetable systems under plasticulture are good examples. In these systems, traditional rotations and soil building practices are usually employed, followed by clean cultivation and the laying of plastic mulch and drip irrigation tape on shaped beds. During the season, large amounts of soluble organic fertilizers — typically fish-based — are fed to the crop through the drip system (*i.e.*, organic fertigation). At the end of the season, all plastics must be removed from the field, and it is returned to more standard organic management.

Ideally, an off-season cover crop will be planted. Such systems are often exceptionally productive and economically attractive, when organic premiums are good. The high cost of soluble organic fertilizer (typically hundreds of dollars/acre), however, plus the marginally higher cost of pest controls, make such systems largely non-competitive in the conventional marketplace. The labeling of such high-input systems as organic presents a paradox for many proponents of organic agriculture.

It is unclear whether these technological advancements reflect the kind of farming most practitioners and supporters of organics think of as truly "organic." To begin with, the research citing environmental and economic benefits has largely been done on low-input organic systems; it is questionable whether similar findings would be made about high-input systems, especially regarding environmental matters. Of particular note, while low-input organic systems are documented as being more resistant to erosion, fields under plastic mulch are reported to be fifteen times more erodible. Traditional organic farms leach minimal amounts of nitrogen into tile or groundwater; the losses from fields loaded with high levels of soluble organic fertilizers is certain to be greater, but how much greater is unknown. The fossil fuel energy involved in plastic manufacture, transportation, and application may or may not be compensated by reductions in tractor fuel use.

Finally, the lowered capital investment required to produce a crop by traditional organic methods makes this form of farming more accessible to resource-poor farmers and entails less risk in years of crop failure or lack of premiums. These factors are less certain in a high-input system. A further consideration is the issue of plastic disposal following removal. At this time, there are few to no options for recycling, and landfills are the fate of plastic mulches at the end of each season.

While it is unwise to rush to judgement regarding high-input organic farming, it is clear that some adaptations will need to be made, if the traditional character and sustainability benefits of organic farming are to be preserved.

Fire

While fire can be used in a number of ways in organic agriculture, the area of greatest interest is *flame* or *thermal weeding*. In its most common application, torches mounted on a tractor toolbar throw a hot flame at the base of mature (*i.e.*, heat-resistant) plants, over the inter-row area, or both. Tractor speed is adjusted so that weeds are not *burned* so much as *seared*. Searing is sufficient to kill most seedling weeds and uses less fuel. Liquid propane (LP) gas is the fuel most commonly used, though alternatives such as alcohol and methane offer the possibility of on-farm sources.

Supplemental Fertilization

In many organic systems, crop rotation, manuring, green manuring, along with enhanced biological activity in the soil, provide an abundant supply of plant-essential minerals annually. This is especially true on naturally deep and rich prairie soils. It is less true on poorer soils and on those that have been heavily exploited through non-sustainable farming practices. To correct mineral deficiencies in organically managed soils, organic growers often apply ground or powdered rock minerals.

The most commonly used rock mineral is high-calcium aglime. Dolomitic limestone, various rock phosphates, gypsum, sulfate of potash-magnesia, and mined potassium sulfate are also common. These are all significant sources of primary (P, K) and/or secondary (Ca, Mg, S) plant nutrients. The savvy organic grower applies significant amounts of these materials only with the guidance of regular soil testing. Less common are other rock powders and fines that are limited sources for the major nutrients but are rich in micronutrients or have some other soil-improving characteristic. Among these are glauconite (greensand), glacial gravel dust, lava sand, Azomite®, granite meal, and others.

Supplementary nutrients that include nitrogen are often provided in the form of animal or plant products and by-products such as fish emulsion, blood meal, feather meal, bone meal, alfalfa meal, and soybean meal. Most of these products also supply some organic matter, though that is not the primary reason they are used.

Evaluating Tools and Practices

One basis on which to evaluate the tools or practices one chooses for an organic operation is whether or not they contribute to biodiversity — biodiversity being one of the principal characteristics of a sustainable organic agriculture.

Crop rotation, cover cropping, farmscaping, companion planting, and intercropping are outstanding examples of practices that contribute to biodiversity. They therefore contribute to the long-term stability and sustainability of the farm agroecosystem. Composting and manuring likewise contribute to biodiversity, but since the diversity they promote is mostly in soil biota, it is rather less obvious to the casual observer.

On the other end of the spectrum are practices such as tillage, cultivation, thermal weeding, solarization, and plastic mulching. Such tools significantly reduce diversity in the field and tend to move the system in a less sustainable direction. This, however, does not necessarily make these practices bad choices.

Organic farming has often been called *natural* farming, as it tries to mimic the processes of nature in producing crops and livestock. However, the analogy goes only so far, since most agricultural systems are characterized by a struggle between human and nature, each with a clear notion of what plants and animals the land *ought* to support and in what proportions. The farmer is typically in the position of "holding back" the natural succession of plant and animal species through the use of diversity-reducing tools and practices. The ideal is to bring about an agricultural system in which the long-term direction emphasizes diversity and sustainability. Among the best visualizations of this ideal are those emerging from the Permaculture movement.

Balanced Nutrition

Whether or not it has been customary in the past, organic growers are encouraged to have periodic soil testing done on their fields. How the results of a soil test are used, however, can vary considerably among farmers, depending on their personal philosophy and management skill. While there is certainly a segment of the organic farming community that has no faith in soil audits, most growers use them as a means to monitor progress in building their soils, to identify nutrient deficits, and to guide supplementary fertilization. While there are many ideas about fertilization guidelines, there are two schools of thought that dominate.

The first is commonly known as the *sufficiency* approach or model. Though somewhat oversimplified, the following are among its significant characteristics:

- annual fertilizer additions of P and K are based in good part on how much the crop is expected to remove from the soil at harvest
- additional amounts of P and K are recommended based on keeping the soil nutrient reserves at a particular level
- lime is added to the soil based on pH
- little to no attention is paid to nutrient balance or to the levels of secondary nutrients Ca and Mg.

The second approach is referred to variously as *cation nutrient balancing*, the *Albrecht system*, the *CEC*, or the *base saturation* approach. It differs from the sufficiency approach in that fertilizer and lime recommendations are made based on an idealized ratio of nutrients in the soil and its capacity to hold those nutrients against leaching.

Cation nutrient balancing is more popular among practitioners of organic farming and sustainable agriculture in general than it is among conventional growers. However, there is no universal agreement on which approach is most appropriate for organic management.

A wide range of other products —humates, humic acids, enzymes, catalyst waters, bioactivators, surfactants, to name a few — are also acceptable in organic crop production. However, the tradeoff between out-of-pocket cost and efficacy of such materials is often challenged by conventional and organic growers alike. Organic growers are encouraged to experiment, but to do so in a manner that allows the actual results to be measured.

Biorational Pesticides

While, in principle, any pesticide use is discouraged in organic systems, a rather wide range of biorational pesticides is permitted. The frequency of pesticide use varies considerably with crop and location. For example, there is virtually no use of pesticides on organic row crops in the Cornbelt. By contrast, organic tree fruits in the Midsouth routinely receive heavy applications of several fungicides and insecticides allowable in organic production. The pesticides permitted in organic farming fall predominantly into several classes.

- *Minerals:* These include sulfur, copper, diatomaceous earth, and clay-based materials like Surround®.

- *Botanicals:* Botanicals include common commercial materials such as rotenone, neem, and pyrethrum. Less common botanicals include quassia, equisetum, and ryania. Tobacco products like Black-Leaf 40® and strichnine are also botanicals but are prohibited in organic production due to their high toxicity.
- *Soaps:* A number of commercial soap-based products are effective as insecticides, herbicides, fungicides, and algicides. *Detergent*-based products are *not* allowed for crop use in organic production.
- *Pheromones:* Pheromones can be used as a means to confuse and disrupt pests during their mating cycles, or to draw them into traps.
- *Biologicals:* One of the fastest-growing areas in pesticide development, biopesticides present some of the greatest hope for organic control of highly destructive pests. Among the most well-known biopesticides are the *Bacillus thuringiensis* (Bt) formulations for control of lepidopterous pests and Colorado potato beetle.

NOTION OF ORGANIC AGRICULTURE

A further notion of organic agriculture that bears addressing is the persistent image of organic farming as being possible only on a very small scale. This impression has been enhanced by the high visibility of organic market gardens. These, of course, are small because market gardening — conventional or organic — is usually done on a smaller scale. Also, some organic market garden systems, such as Biointensive Mini-Farming, use highly labour intensive/low capital investment technologies. These have become popular among U.S. gardeners and, more importantly, with those concerned with Third World development, where such systems are especially relevant. Focus on these systems has, unfortunately, distorted the picture of organics as a whole. Traditionally, organic farms truly have been smaller than conventional operations. This has been due in part to labour requirements. Organic systems are generally more labour intensive. Studies done in the late 1970s by Washington University, for example, found that about 11% more labour was required per unit of production where agronomic crops were concerned.

This difference can be much greater where horticultural crops are involved, and farm size may be limited accordingly. However, technological innovations in organic horticultural production are helping to narrow the gap. Organic systems are also more information intensive, requiring additional management time in planning, pest scouting, and related activities. For this reason, organic management can be better done if a farm is not too large. Essentially, the notion that organic systems are only possible on very small farms is a false one. Both the Washington University and the USDA studies confirmed this. Given the range of acceptable technologies available, organic agriculture can be sized to fit a wide range of farms and enterprises.

LANDMARK RESEARCH

Throughout its early history, organic agriculture was treated with either hostility or apathy by the USDA, land grant universities, and conventional agriculture in general. Fortunately, the atmosphere for discussing and investigating organics has improved considerably. While it did not become boldly evident until the 1990s, the tide actually began turning in the late 1970s and early 1980s.

A number of factors precipitated this change, among them the growth in the organic industry. Serious money demands seri-ous attention. Also critical, from the perspective of the research community especially, were some landmark studies that lend credibility to organic farming as a truly viable option for American agriculture. The first of these "landmarks" was a series of studies done by Washington University.

Funded by the National Science Foun-dation, this research was motivated by the energy crisis of the 1970s and the effect that higher energy prices would have on agriculture in the nation's Cornbelt. When researchers learned that there were commercial farms that were not dependent on the high-energy inputs of conventional farming, the focus quickly shifted to the study of organics.

In addition to the documentation of practices, crop yields, attitudes, and the sustainability indices (cited elsewhere in this publication), the researchers made what was certainly the most astounding discovery of all, that commercial organic farms could be competitive with conventional farms in the conventional marketplace. Arriving on the heels of the Washington University work was another study of great significance done by the USDA. In contrast to the Washington University effort, these researchers chose to extend their survey of farmers nationally and over a wide range of enterprises.

The findings of the USDA study, which were fair, largely positive, and encouraging, kicked open the door for future organic research in ways that a non-land grant/non-USDA entity like Washington University could not. The final report – bound with pastel green cover sheets — was a conspicuous object at alternative agriculture conferences and field days throughout the early 1980s

Also of particular note was a symposium on organic farming held in Atlanta, Georgia, in late 1981. The meeting was spon-sored jointly by the American Society of Agronomy, the Crop Science Society of America, and the Soil Science Society of America — traditionally very conservative entities. It brought together not only representatives of the Washington University and USDA teams, but a surprising number of other researchers clearly interested in the same issues of sustainability and finding a glimmer of hope in organic agriculture.

Organic Certification

In 2002, when the USDA adopted the National Organic Standard that spells out what farmers and food processors must (and must not) do to be certified "organic," the organic industry already had a long history of relying on third-party certifiers to ensure the integrity of their products and practices.

Under this system, a state-run or accredited private agency (the third-party) evaluates farmers and processors to see whether they conform to the standards of the National Organic Programme (NOP). Those who do can then market their products as "USDA Certified Organic" and display the official USDA organic seal on their packaging.

In essence, certification is largely about integrity — assuring that the buyer is getting what he or she is paying for. Certified organic production, then, means production by approved organic methods, with additional pains taken to eliminate contamination with prohibited materials and commingling with conventional products. There is a common misconception that certified organic means "pesticide residue-free." Consumers have a right to expect little or no pesticide residue on certified organic crops because none are used in their production. However, ours is a dirty world in which pesticides and their break-down products are ubiquitous. This is only to be expected in a national farm system where more than 99% of all applied farm chemicals miss the target organism.

The principles, practices, and tools discussed in the remainder of this publication reflect the guidelines recognized by the NOP, though minor details may vary among third-party certifiers. It is important, therefore, that producers understand their certifying agency's standards well and keep in close touch with a representative. Note that these principles and practices also provide a foundation for other sustainable approaches to crop production, perennial or annual.

ORGANIC PRINCIPLES

There are several compelling principles that characterize certified organic farming. They include *biodiversity, integration, sustainability, natural plant nutrition, natural pest management*, and *integrity*. Most organic operations will reflect all of these to a greater or lesser degree. Since each farm is a distinct entity, there is a large degree of variation.

BIODIVERSITY

As a general rule, diverse ecosystems in nature have a higher degree of stability than those with only a few species. The same is essentially true for agroecosystems. Farms with a diverse mix of crops have a better chance of supporting beneficial organisms that assist in pollination and pest management. Diversity above ground also suggests diversity in the soil, providing better nutrient cycling, disease suppression, tilth, and nitrogen fixation. Good organic farmers mimic the biodiversity of nature through practices like intercropping, companion planting, establishment of beneficial habitats, and crop rotation (sometimes referred to as companion planting across time). The effort to increase biodiversity works hand-in-hand with enterprise diversity, which is often (but not necessarily) an objective on organic farms.

Diversification and Integration of Enterprises

The drive to build biodiversity in organic systems encourages diversity among enterprises, but not as isolated or independent entities. Good organic operations integrate their various enterprises. A good example can be seen among Midwestern organic crop and livestock operations. The typical Midwestern organic operation ties the needs of crops and livestock together in a practi-cal and elegant way. The forage and grain needs of ruminant livestock make for a diverse mix of crops. Particularly valuable is the inclusion of le-gume forages for ruminant feed. Forage legumes in rotation fix a sustainable supply of nitrogen in the soil that feeds subsequent non-legume crops in rotation. Manure from the livestock enterprises is conserved as a nutrient resource and recycled back to the crop fields. Farms such as these have the additional advantage of greater economic sustainability, as their risks are spread over several livestock and crop enterprises.

Sustainability

In addition to the greater economic sustainability afforded by enterprise diversification, organic farmers are often able to reap market premiums for certified production. However, since many organ-ic enterprises realize somewhat lower marketable yields, this has not always translated into higher profits or greater economic sustainability. As more and more organic growers enter the marketplace, it is likely that premiums will stabilize at modest levels and may vanish for some crops. Organic producers need to look well ahead and be aware of shifting trends.

As alluded to earlier, many U.S. organic farms per-form well on many of the measurable indicators associated with sustainability, such as energy con-sumption and environmental protection. However, sustainability is an ideal, and the best that can be said is that current organic farms are closer to the ideal than most alternatives—certainly closer than comparable conventional farming operations.

Index